Friction Stir Welding of High-Strength 7XXX Aluminum Alloys

Friction Stir Welding of High-Strength 7XXX Aluminum Alloys

Rajiv S. Mishra and Mageshwari Komarasamy
Department of Materials Science and Engineering,
University of North Texas, Denton, TX, USA

AMSTERDAM • BOSTON • HEIDELBERG • LONDON
NEW YORK • OXFORD • PARIS • SAN DIEGO
SAN FRANCISCO • SINGAPORE • SYDNEY • TOKYO
Butterworth-Heinemann is an imprint of Elsevier

ELSEVIER

Butterworth-Heinemann is an imprint of Elsevier
The Boulevard, Langford Lane, Kidlington, Oxford OX5 1GB, UK
50 Hampshire Street, 5th Floor, Cambridge, MA 02139, USA

Library of Congress Cataloging-in-Publication Data
A catalog record for this book is available from the Library of Congress

British Library Cataloguing-in-Publication Data
A catalogue record for this book is available from the British Library

ISBN: 978-0-12-809465-5

For Information on all Butterworth-Heinemann publications
visit our website at http://www.elsevier.com/

Working together
to grow libraries in
developing countries

www.elsevier.com • www.bookaid.org

Publisher: Joe Hayton
Acquisition Editor: Christina Gifford
Editorial Project Manager: Heather Cain
Production Project Manager: Anusha Sambamoorthy
Cover Designer: MPS

Typeset by MPS Limited, Chennai, India

CONTENTS

List of Figures...vii
List of Tables.. xiii
Preface to This Volume of *Friction Stir Welding and*
Processing Book Series...xv

Chapter 1 Introduction 1
References...3

Chapter 2 Physical Metallurgy of 7XXX Alloys................................5
2.1   Precipitation Reaction in Al-Zn-Mg Alloys.................................5
2.2   Effect of Cu..8
2.3   Effect of Ag..9
2.4   Effect of Microalloying...10
2.5   Effect of Li..10
2.6   Effect of Predeformation on Aging..11
2.7   Summary...11
2.8   Differential Scanning Calorimetry Observations 12
References...13

Chapter 3 Friction Stir Welding—Overview 15
3.1   Introduction..15
3.2   Taxonomy 15
3.3   Various Zones...16
3.4   Material Flow...17
3.5   Defects in FSW 18
3.6   Key Benefits of FSW 19
References...20

Chapter 4 Temperature Distribution 21
4.1   Introduction..21
4.2   Experimental Observations...21

4.3   Numerical Observations 24
4.4   Summary 26
References...27

Chapter 5 Microstructural Evolution 29
5.1   Introduction..29
5.2   Evolution of Grain Size...29
5.3   Precipitate Evolution 33
5.4   Differential Scanning Calorimetry...........................38
5.5   Summary 45
References...46

Chapter 6 Mechanical Properties....................................49
6.1   Introduction..49
6.2   Hardness and Tensile Properties..............................49
6.3   Fatigue and Damage Tolerance................................70
6.4   Joint Efficiency 84
References...88

Chapter 7 Corrosion 91
7.1   Introduction..91
References...99

**Chapter 8 Physical Metallurgy-Based Guidelines for
             Obtaining High Joint Efficiency........................ 101**

Chapter 9 Summary and Future Outlook........................ 103

LIST OF FIGURES

Figure 2.1   Summary of various reactions in Al-Zn-Mg-based   12
             alloys
Figure 3.1   Schematic illustration of the FSW process   16
Figure 3.2   A macrograph of the cross-section showing   17
             various regions in FSP 7075-T651
Figure 3.3   Various defects in FSW welds   19
Figure 4.1   Temperature distribution adjacent to the FSW   22
             nugget of 7075-T651
Figure 4.2   (A) Location of the thermocouples and (B) peak   23
             temperature variation in three locations as a
             function of the welding speed
Figure 4.3   (A) Torque and peak temperature variation as   23
             a function of rotation rate under two conditions,
             (B) temperature–time at 8 ipm, and (C) time spent
             above 200°C in HAZ under all the three
             conditions
Figure 4.4   Temperature–time profiles (A) for welds with pitch   25
             0.28 mm/rev and (B) for welds with 180 rpm
Figure 5.1   TEM micrographs of displaying the grain   32
             structure in different weld zones, (A,B) base metal,
             (C) HAZ, (D) TMAZ I, (E) TMAZ II, and
             (F) DXZ
Figure 5.2   SAXS maps (volume fraction and size) of η 35
             precipitates in various zones for the T3 and T79
             welds under low and high welding speeds
Figure 5.3   Postweld microstructural characterization of   37
             different weld zones, grain boundary precipitates
             in (A) nugget, (D) TMAZ, (G) HAZ; bimodal
             precipitates in (B) nugget, (E) TMAZ;
             fine strengthening precipitates in (C) nugget,
             (F) TMAZ; (H) coarsened HAZ precipitates

Figure 5.4   Precipitate structure in (A) base material, (B) HAZ,   39
              (C) TMAZ I, (D) TMAZ II, and (E) DXZ
Figure 5.5   Precipitate structure in TMAZ (A) arranged like a   40
              deformed grain boundary and (B) preferential
              precipitation on subgrain boundaries
Figure 5.6   Overall summary of the precipitate evolution in   40
              different zones under various conditions
Figure 5.7   A typical DSC thermogram with baseline   41
              correction in the case of solution-treated and
              solution-quenched sample
Figure 5.8   DSC peaks of the HAZ region of 7050-T7651   43
              friction stir weld as a function of the natural aging
              time
Figure 5.9   DSC peaks of the weld nugget region of 7050-   43
              T7651 friction stir weld as a function of the
              natural aging time
Figure 5.10   DSC peaks of weld center, advancing, and   44
              retreating sides of the 400-rpm weld
Figure 6.1   Weld thermal cycle measured at minimum HAZ   50
              hardness and the corresponding schematic of the
              precipitate evolution
Figure 6.2   Local strain distribution in nugget and HAZ   51
              regions
Figure 6.3   Hardness profiles across the weld at (A) weld top   52
              and root in as-FSW condition, (B) weld top in
              as-SFW+T6, and (C) weld root in
              as-FSW and as-FSW+T6
Figure 6.4   (A) As-welded hardness across the weld as a   53
              function of the weld speed for an advance per
              revolution of 0.42 mm/rev and (B) average nugget
              hardness as a function of welding speed for three
              advance per revolutions
Figure 6.5   Variation in average minimum HAZ hardness   53
              as a function of the welding speed in (A) as-welded
              and (B) postweld heat-treated condition
Figure 6.6   Change in hardness in nugget and HAZ in   54
              response to postweld heat treatment as a function
              of the peak temperature

Figure 6.7   (A) Hardness profile across the weld as a function   55
              of the tool rotation rate or spindle speed and
              (B) effect of tool rotation rate on the hardness
              variation for the three tool traverse speeds
Figure 6.8   (A) Weld cross section, (B) hardness contour map,   56
              microhardness as a function of (C) tool rotation
              rate, (D) traverse speed, and (E) rev/min
Figure 6.9   Postweld heat-treated hardness profile across the   57
              weld in (A) 800 rpm and 16 ipm and (B) 200 rpm
              and 6 ipm
Figure 6.10   Hardness distribution across the weld under   57
              various conditions
Figure 6.11   Hardness measurements across the weld mid-plane   58
              in W, T6, and T7 tempers after the postweld heat
              treatment
Figure 6.12   (A) Digital image correlation analysis of strain   59
              evolution across the weld during transverse-weld
              tensile test and (B) transverse-weld tensile test
              results of the base metals and the welds after
              postweld heat treatment along with the DIC
              stress–strain results
Figure 6.13   Across the weld hardness measurements of the   60
              friction stir welds in different initial base metal
              tempers
Figure 6.14   Microhardness of the actively cooled weld in   62
              as-welded and 1000 h naturally aged conditions
Figure 6.15   Tensile test results of the FSW 7075-T351   63
              (A) yield and tensile strength and (B) elongation
Figure 6.16   Hardness evolution across the weld as a function   64
              of natural aging times in (A) 7050-T7651 and
              (B) 7075-T651 Al alloys
Figure 6.17   Tensile properties of (A) 7050-T7651 and   65
              (B) 7075-T651 Al alloys as a function of the
              natural aging time
Figure 6.18   Mid-plane hardness across the weld in as-welded   66
              and naturally aged (3 and 6 years) specimens
Figure 6.19   Transverse-weld hardness profiles of all the five   67
              passes

Figure 6.20   Transverse-weld hardness profile at three   68
               depth from the top surface in (A) as-welded,
               (B) postweld heat-treated conditions, and
               (C) variation in the weld centerline hardness as a
               function of the distance from the weld root in
               aged condition
Figure 6.21   Transverse-weld profiles in as-welded condition at   69
               various depths from the top surface
Figure 6.22   Transverse-weld hardness profiles in as-welded   70
               condition and T7 aged condition at 2 mm from the
               top surface
Figure 6.23   (A) Engineering stress–strain curves, and (B) S–N   71
               curves for base metals and FSW joints in various
               initial tempers
Figure 6.24   S–N curves for base metal, flawless FS welds, and   72
               welds with kissing bonds of 315 and 670 μm along
               with the inset showing two failure locations
               in 670-μm kissing bond
Figure 6.25   Observation and the explanation of the crack   73
               growth process
Figure 6.26   Microhardness contour maps of (A) 800 rpm   74
               and 4 ipm, (B) 800 rpm and 8 ipm, (C) 800 rpm
               and 16 ipm, (D), 1000 rpm and 16 ipm, and
               (E) 1200 rpm and 16 ipm
Figure 6.27   (A) Macroscopic view of the failed sample along   75
               with the crack, (B) overall view of the fracture
               surface with multiple crack initiation sites,
               (C) and (D) high magnification images of the
               location 1 which were close to the TMAZ region,
               and (E) location 2 initiation site which was along
               the nugget
Figure 6.28   S–N curves for the base metal and the FSW joint   76
Figure 6.29   Survey of fatigue failure locations in FSW joints   77
Figure 6.30   Fatigue crack growth behavior of nugget, HAZ   78
               and parent material, tested at $R = 0.33$ and
               $R = 0.7$ stress ratios

Figure 6.31   (A) Residual stress measurements on the ESE(T)   79
specimen, residual stress components in
(B) longitudinal, and (C) transverse directions

Figure 6.32   Fatigue crack growth behavior of base metal and   80
HAZ region

Figure 6.33   Variation in crack length with number of fatigue   81
cycles for various samples

Figure 6.34   Residual stress mapping across the weld top   82
surface in as-welded and peened conditions in
(A) longitudinal and (B) transverse orientations

Figure 6.35   Crack length versus number of fatigue cycle in two   83
stress ratios (A) $R = 0.1$ and (B) $R = 0.7$

Figure 7.1   Exfoliation test results after various postweld heat   92
treatments

Figure 7.2   Immersion study to observe the distinct response   94
of the (A) heat-affected zone and (B) nugget
regions to the corrosive environment (Images
acquired in BSE mode)

Figure 7.3   Corrosion response to 3 wt.% NaCl solution in   94
(A) thermomechanically affected zone and
(B) heat-affected zone

Figure 7.4   Grain boundary sensitization in the   95
thermomechanically affected region of a
7050-T7451 weld

Figure 7.5   Pitting potentials in various zones of   95
(A) 7075-T651 and (B) 7050-T7451 FSW

Figure 7.6   Observation of the corrosion products after the   96
immersion testing in (A) 7075-T651 and
(B) 7050-T7451 FSW

Figure 7.7   Transverse cross section observation after the   98
immersion testing in (A) 7075-O FSW, (B) 7075-
T7 FSW, and (C) T7 postweld heat-treated 7075-O
FSW

Figure 8.1   An illustration of scaling of tool with the thickness   102
of material. It changes the angle of HAZ and the
associated fracture path during transverse loading

Figure 8.2   An illustration of single pass versus double pass   102
              approach for thick plates. Note the change in the
              overall HAZ path in the double pass weld. The
              associated fracture path during transverse loading
              will change as the crack path becomes more
              complicated for the double pass weld

LIST OF TABLES

Table 2.1   Composition of Selected 7000 Alloys Starting From   6
             Simple Al-Zn System to Al-Zn-Mg, Al-Zn-Mg-Cu,
             and Al-Zn-Mg-Cu-Ag

Table 2.2   Stoichiometry, Orientation of Precipitates, and Plane   7
             of Preferential Formation of the Various Phases in
             the Precipitation Sequence of 7000 Series Aluminum
             Alloys

Table 2.3   Effect of Cu on the Precipitate Evolution   8

Table 2.4   Precipitation Sequence as a Function of Time   9
             at 121°C

Table 2.5   Effect of Alloy Composition on the DSC Dissolution   13
             Peak of η Precipitates

Table 4.1   Summary of Peak Temperatures Observed in Various   26
             Weld Zones

Table 5.1   Literature Information Available on the Weld Nugget   33

Table 5.2   Literature Information Available on FSW HAZ   34

Table 5.3   Summarization of the Microstructural Evolution   38
             in Different Zones in 7449 TAF FSW Welds

Table 5.4   Summary of Precipitate Evolution in HAZ and   40
             Nugget as a Function of Natural Aging Time

Table 5.5   Overall Summary of a Few Observations of   45
             Microstructural Evolution in 7XXX Alloys

Table 6.1   Summary of the Longitudinal Nugget Tensile   50
             Properties of Friction Stir Welded 7075-T651
             Al Alloy

Table 6.2   Summary of the Transverse-Weld Tensile Properties   50
             of Friction Stir Welded 7075-T651 Al Alloy

Table 6.3   Summary of the Tensile Properties of the 7050-T7451   52
             FSW Welds

Table 6.4   Summary of the Transverse-Weld Tensile Test Results   60
             Along With Failure Location

Table 6.5   Tensile Properties of Base Metal and the Processed   61
             Material
Table 6.6   Summary of the Tensile Results as a Function of the   66
             Natural Aging Time
Table 6.7   Summary of the Fatigue Crack Initiation Locations   73
             Under Various Conditions
Table 6.8   Minimum HAZ Hardness–Based Joint Efficiency for   84
             Various 7XXX Alloys Under Different FSW
             Conditions
Table 6.9   Transverse Ultimate Tensile Strength–Based Joint   86
             Efficiency for Various 7XXX Alloys Under Different
             FSW Conditions
Table 7.1   Summary of the Exfoliation Test Ratings After   93
             Various Postweld Heat Treatments in 7075-T73 Weld
Table 7.2   Summary of the Exfoliation Test Ratings After Various   93
             Postweld Heat Treatments in 7075-T6 Weld
Table 7.3   A Detailed Summary of the Microstructural Features   98
             at Various Zones, the Expected Corrosion Response,
             and the Combination of the Appropriate to
             Technique to Characterize the Localized Corrosion
             Response

PREFACE TO THIS VOLUME OF *FRICTION STIR WELDING AND PROCESSING* BOOK SERIES

This is the sixth volume in the recently launched short book series on *Friction stir welding and processing*. As highlighted in the preface of the first book, the intention of this book series is to serve engineers and researchers engaged in advanced and innovative manufacturing techniques. Friction stir welding was invented more than 20 years back as a solid-state joining technique. In this period friction stir welding has found a wide range of applications in joining of aluminum alloys. Although the fundamentals have not kept pace in all aspects, there is a tremendous wealth of information in the large volume of papers published in journals and proceedings. Recent publications of several books and review articles have furthered the dissemination of information.

This book is focused on friction stir welding of 7XXX alloys, a topic of great interest for practitioners of this technology in the aerospace sector. 7XXX series alloys are among the highest specific strength aluminum alloys, and achieving high joint efficiency is a desired goal. The basic research toward this has reached a level that warrants compilation of the scientific knowledge. This volume provides a good summary of the current understanding and provides brief guideline for future research directions. It is intended to serve as a resource for both researchers and engineers dealing with the development of high-efficiency structures. As stated in the previous volumes, this short book series on friction stir welding and processing will include books that advance both the science and technology.

Rajiv S. Mishra
University of North Texas, Denton, TX, United States

April 23, 2016

CHAPTER 1

Introduction

The aerospace industry has been aggressively pushing the limits of the high-strength aluminum alloys in strength, damage tolerance, and corrosion resistance fronts to develop strong and tough aluminum alloys for various parts to increase the overall efficiency. Attempts were made by tweaking the solute composition, modifying the heat treatment conditions for specific requirements, and/or with the additions of the microalloying elements such as Ag and Sc. These endeavors have been extremely successful in producing various high-strength Al alloys with optimized properties. A thorough discussion on the physical metallurgy of various 7XXX alloys are given in Chapter 2, Physical Metallurgy of 7XXX Alloys. This chapter completely focuses on the precipitation sequence in various 7XXX alloys, and the effect of composition and temperature on these sequences. A major concern is the joinability of these high-strength alloys as these alloys have been regarded as unweldable through the conventional fusion-welding techniques. Fusion welding of high-strength Al alloys produced intermetallics, numerous solidification defects, and distortion of the components. These factors resulted in a very low joint efficiency of the high-strength Al fusion welds that made it inappropriate for structural components. Mechanical fastening, like riveting, has been most popular solution for joining of these high-strength aluminum alloys. The large quantity of rivets used in airplane significantly increases the weight of the plane along with the added complexity of stress concentration and corrosion.

The problems with welding of these high-strength Al alloys and the weight penalty associated with rivet joining changed dramatically with the invention of a solid-state welding method. Thomas et al. [1] invented the solid-state welding technique, commonly known as friction stir welding (FSW) at The Welding Institute of United Kingdom in 1991. This invention revolutionized the method of joining the aluminum alloys for airplane, fuel tanks of rockets and space shuttles, automotives, ships, rails, and numerous other structural components [2]. Enormous research

Friction Stir Welding of High-Strength 7XXX Aluminum Alloys. DOI: http://dx.doi.org/10.1016/B978-0-12-809465-5.00001-5

has been conducted toward the basic understanding of the new solid-state welding process, by both academic and industrial researchers. This is one of the fastest adopted welding technology by the most stringent industry focused on property requirements, the aerospace industry.

During the FSW in butt configuration, the plates to be joined are clamped tightly. A nonconsumable rotating tool with pin and shoulder is inserted into the abutting faces of the plates and after reaching the predetermined plunge depth (determined by the height of tool pin), the tool traverses along the abutted line. The material around the rotating tool does not reach the melting point. The rotating FSW tool softens or plasticizes the material so that it can be moved around the tool and subsequently deposited at the back of the moving tool. Description of the FSW process is given in Chapter 3, Friction Stir Welding—Overview, and for more information the reader can refer to the recent book on FSW and processing [2]. Generally, the temperature in the region where the stirring action happens reaches above 400°C for aluminum alloys [3]. This is extremely high temperature for the strengthening precipitates in 7XXX alloys. The welded region undergoes precipitate coarsening, dissolution, and reprecipitation during the FSW thermomechanical processing. The presence of large strain and high temperature in the welded region results in the formation of dynamically recrystallized grains. Furthermore, like fusion-welding technique, FSW also exhibits a temperature distribution from the edge of the stirred volume toward the base metal. Due to this distribution in temperature, a variety of precipitate evolution can be expected. The compilation of both experimental measurements of temperature from various regions in the FSW joint and numerical simulations is provided in Chapter 4, Temperature Distribution. As mentioned earlier, the high-temperature severe plastic deformation leads to a variation in microstructural evolution across the various regions, which is covered in Chapter 5, Microstructural Evolution [4]. The microstructural characteristics, such as grain size, precipitate size, number density, and distribution, and dislocation density, control the mechanical properties. The FSW welds intrinsically possess temperature variation and corresponding microstructural variation across the weld, which in turn determines the final properties across the weld. As in all welding processes, the FSW joint is not a monolithic structure in terms of properties. The properties vary quite significantly across the weld joint. Chapter 6, Mechanical Properties, is devoted to the analysis of various mechanical properties, such as hardness variation across the weld, joint tensile properties, fatigue and damage tolerance, and a compilation of the weld joint efficiencies. In the case of any structural components, the weakest part controls the overall life and eventually leads to the failure. In the case of the FSW joint, the weakest point in majority of the cases lies along the heat-affected zone (HAZ) or the HAZ–thermo-mechanically affected zone boundary. The micro-structural characteristics also control the corrosion property, and given the uniqueness, the corrosion resistance also varies among various zones, which is discussed in Chapter 7, Corrosion. Given the overall knowledge and the compilation of various aspects, this short book finishes with a discussion on the process and alloy considerations for better FSW 7XXX welds. Overall, the scope of this short book is to provide a broad overview of various aspects of the friction stir welds.

REFERENCES

[1] W.M. Thomas, E.D. Nicholas, J.C. Needham, M.G. Murch, P. Templesmith, C.J. Dawes, G.B. Patent Application No. 9125978.8 (December 1991).

[2] R.S. Mishra, P.S. De, N. Kumar, Friction Stir Welding and Processing, Springer International publishing, Switzerland, 2014.

[3] Kh.A.A. Hassan, P.B. Prangnell, A.F. Norman, D.A. Price, S.W. Williams, Effect of welding parameters on nugget zone microstructure and properties in high strength aluminium alloy friction stir welds, Sci. Technol. Weld. Join. 8 (2003) 257–268.

[4] J.-Q. Su, T.W. Nelson, R. Mishra, M. Mahoney, Microstructural investigation of friction stir welded 7050-T651 aluminium, Acta Mater. 51 (2003) 713–729.

CHAPTER 2

Physical Metallurgy of 7XXX Alloys

The interesting properties of the 7000 series Al alloys mainly stem from the presence of the strengthening precipitates. These precipitates form from the supersaturated solid solution upon various aging heat treatments. A selected list of various 7000 series alloys are given in Table 2.1. It contains the examples of Al-Zn, Al-Zn-Mg, Al-Zn-Mg-Cu, and Al-Zn-Mg-Ag(Cu) alloys. The Zn:Mg ratio of around 3 is mostly preferred to avoid the formation of S (Al_2CuMg) and T ($Mg_3Zn_3Al_2$) phases.

A general and most accepted sequence of precipitation in various 7000 alloys can be represented as:

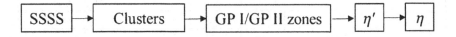

$$\boxed{\text{SSSS}} \rightarrow \boxed{\text{Clusters}} \rightarrow \boxed{\text{GP I/GP II zones}} \rightarrow \boxed{\eta'} \rightarrow \boxed{\eta}$$

During quenching or immediately after, the clusters form rapidly. Depending upon the alloy composition, the details of the solute or solute–vacancy clusters would vary which will be discussed in detail in Section 2.1. First, various phases in Al-Zn-Mg alloys will be discussed followed by Cu and Ag containing Al-Zn-Mg alloys. The atomic size difference impacts the lattice strain and the type of precipitates, although other factors also play important role.

2.1 PRECIPITATION REACTION IN Al-Zn-Mg ALLOYS

In Al-Zn-Mg alloys, Zn-vacancy clusters and Mg aggregates form immediately after quenching. Guinier–Preston (GP) zones form after months of natural aging [3]. When the naturally aged material is heat treated at 95°C, the reversion of Mg-rich clusters and the retention of Zn-vacancy clusters have been observed. When heat treatment temperature is raised to 150°C, a direct formation of semicoherent or incoherent precipitates and also the retention of Zn-vacancy clusters

Friction Stir Welding of High-Strength 7XXX Aluminum Alloys. DOI: http://dx.doi.org/10.1016/B978-0-12-809465-5.00002-7

Table 2.1 Composition of Selected 7000 Alloys Starting From Simple Al-Zn System to Al-Zn-Mg, Al-Zn-Mg-Cu, and Al-Zn-Mg-Cu-Ag [1,2]						
	Alloy Designation	Zn	Mg	Cu	Ag	Zn(+ Cu):Mg
Al-Zn alloys	7072	0.8–1.3	0.10 max	0.10 max	–	–
Al-Zn-Mg alloys	7039	3.5–4.5	2.3–3.3	0.10 max	–	1.4
	7005	4.0–5.0	1.0–1.8	0.10 max	–	3.2
	7020	4.0–5.0	1.0–1.4	0.20	–	3.8
	7008	4.5–5.5	0.7–1.4	0.05 max	–	4.8
Al-Zn-Mg-Cu alloys	7079	3.8–4.8	2.9–3.7	0.40–0.8	–	1.5
	7022	4.3–5.2	2.6–3.7	0.50–1.00	–	1.8
	7075	5.1–6.1	2.1–2.9	1.2–2.0	–	2.9
	7001	6.8–8.0	2.6–3.4	1.6–2.6	–	3.2
	7010	5.7–6.7	2.2–2.7	1.5–2.0	–	3.3
	7178	6.7–7.3	2.4–3.1	1.6–2.4	–	3.3
	7100	6.12	2.25	1.71	–	3.5
	7012	5.8–6.5	1.8–2.2	0.8–1.2	–	3.6
	7150	5.7–6.7	1.8–2.7	1.5–2.3	–	3.6
	7050	5.7–6.7	1.9–2.6	2.0–2.6	–	3.8
	7049	7.2–8.2	2.0–2.9	1.2–1.9	–	3.8
	7040	5.7–6.7	1.7–2.4	1.5–2.3	–	4.0
	7449	7.5–8.7	1.8–2.7	1.4–2.1	–	4.4
	7055	7.6–8.4	1.8–2.3	2.0–2.6	–	5.0
	7085	7.0–8.0	1.2–1.8	1.3–2.0	–	6.1
Al-Zn-Mg-Cu-Ag alloys	7009	5.5–6.5	2.1–2.9	0.6–1.3	0.25–0.40	2.8
	7047	7.0–8.0	1.3–1.8	0.04	0.20–0.50	4.9

during the initial 30 min have been observed. The GP zones form on these remaining, highly stable Zn-rich clusters [3]. Both GP I and GP II zones were observed in the following heat treatment conditions: 1.5–5 h at 100°C [4,5] and 1 h at 130°C [6]. Even after 7 h at 120°C, only GP II zones were observed and there was no observation of the metastable η' [6]. The following observations were made after 5 h at 100°C and 6 h at 150°C heat treatments: dissolution of GP I zones and the presence of GP II, η', and η phases [5]. In the case of GP zones, an equal fraction of Mg and Zn content was observed and this was due to the size difference of these atoms. Most likely, η may be nucleated on η' rather than the dissolution and reprecipitation process [7]. As can be clearly noted, the type of phase formation and its retention highly depends on the aging conditions. Higher Zn content increases the precipitate number density, allows homogeneous nucleation, and reduces the precipitate size [8]. Moreover it also plays a role in reducing the quench sensitivity. The description of various precipitate phases is given in Table 2.2.

Table 2.2 Stoichiometry, Orientation of Precipitates, and Plane of Preferential Formation of the Various Phases in the Precipitation Sequence of 7000 Series Aluminum Alloys [4,9–11]	
Phases	**Remarks**
Clusters	• Both solute (Zn, Mg, Cu) and vacancy-rich solute clusters have been observed prior to the formation of Guinier–Preston (GP) zones.
GP I zones	• Ordered and coherent layers of Zn and Mg/Al in {100} Al planes that mostly exhibit spherical morphology. • Serve as a precursor for the precipitation of η'. • Presence of Cu in GP I zones was observed and believed to have been involved in the initial clustering process. • Forms over a wide range of temperature, from RT to 140–150°C. The formation of GP I zones is closely related to the solute clusters.
GP II zones	• The precursors are the vacancy-rich solute clusters. Al, Zn, and Mg and/or Cu atoms on the {111} Al planes that exhibit plate-like morphology and fully coherent with the matrix. • Typical GP II zones sizes are 1–6 atomic layer in thickness along {111} Al planes and 3–6 nm in width. Transform directly to the η' precipitate. More thermally stable than GP I zones. • GP II zone formation was observed when quenched from 450°C. The formation is highly dependent on the quenched in vacancies.
η'	• η', Mg (Zn, Al, Cu)$_2$, precipitates on the {111} Al planes and possesses a hexagonal close-packed crystal structure with the lattice parameters $a = 0.496$ nm and $c = 6d_{111Al} = 1.402$ nm. They are coherent with the matrix and exhibits well-defined crystallographic relationship with the Al matrix, $(0001)_{\eta'}//\{111\}Al$, $[11\bar{2}0]_{\eta'}// < 112 > Al$, $(10\bar{1}0)_{\eta'}//(110)_{Al}$. • η' is the metastable phase that is responsible for peak strength and is found in T6 condition. Typical sizes of η' precipitates are of 3–4 nm in thickness and 5–10 nm in width. These are plate-like precipitates. The observed Zn:Mg ratio was 1:1 as opposed to the expected 1:2 ratio [8].
η Precursor [11]	• Present in Cu containing Al-Zn-Mg alloys. • The lattice parameters are $a = 0.496$ nm and $c = 4d_{111Al} = 0.935$ nm with the chemistry of Mg (Zn, Al, Cu)$_2$. Orientation relationship with Al matrix is same as η' and they are also coherent with the matrix. This is also considered as the hardening phase. • This is mostly considered an intermediate phase that aids the η' to η transformation process. Note that a and c lattice parameter of the η precursor is same as the η' and η, respectively.
η	• Precipitates are mostly of MgZn$_2$ type with very little Al and are incoherent with the matrix. The lattice parameters are $a = 0.521$ nm and $c = 0.860$ nm. • Unlike the η' and η precursor, this phase is known to exhibit multiple orientation relationship with Al matrix. • These are cigar-like precipitates and are found in T7 condition. The observed Zn:Mg ratio is close to 1:1.6 as opposed to the expected 1:2 ratio.

Table 2.3 Effect of Cu on the Precipitate Evolution [13]	
Cu Content	Main Precipitate Type After 30 min at 140°C
0.01	GP I
2.04	GP I + GP II
2.98	η' + GP II
3.92	η' + GP II

2.2 EFFECT OF Cu

Cu produces multifold effects on the precipitation characteristics in various 7000 alloys. Cu enhances both the rate and the extent of hardening. Both hardness and tensile properties have increased in Cu containing Al-Zn-Mg alloys. Up to 2.5 wt.% of Cu can be dissolved into η' to η precipitates and S phase (Al$_2$CuMg) forms above that limit when heat treated above 175°C. In high Cu content alloys, intermetallic S phase forms during the solidification process [12]. Due to the strong interaction with vacancies and other solute atoms (Zn and Mg), the presence of Cu promotes an early clustering process [10,13,14]. Cu also stimulates the formation of GP II zones from the clusters and its subsequent transformation to η' [13]. The effect of Cu on the early precipitate evolution in Al-6.8 wt.% Zn-2.8 wt.% Mg is shown in Table 2.3. The alloy without Cu had only GP I zones and, on the other hand, η' precipitates formed in the alloys containing Cu [13].

Presence of even low amount of Cu (0.13 at.%) decelerated the GP I zone dissolution during 150°C thermal exposure which otherwise dissolves completely [15]. Cu partly modifies the GP zones shape from spherical to ellipsoidal. Furthermore, it retards the transformation of η' to η, thereby maintaining the peak strength for longer durations [13]. An intermediate phase between η' and η is known to form and is called as η precursor [11]. The diffusivity of Cu in Al is lower than the Mg and Zn diffusivity in Al. Therefore, the precipitate coarsening has been retarded in the presence of Cu [13,15].

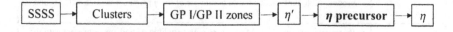

Presence of Cu in Al-Zn-Mg alloys reduces the stress corrosion cracking (SCC) susceptibility, but increases the quench sensitivity and also reduces the weldability. In 7050 Al alloys (Al-Zn-Mg-Cu), Sha et al. [10] analyzed

Table 2.4 Precipitation Sequence as a Function of Time at 121°C [10]	
Aging Time (h)	Observations
0.5	• High density of GP I zones with wide distribution in size, spherical to blocky to elongated (along <110>). • GP I zones nucleated homogeneously from the supersaturated solid solution. • GP I zones contain Zn, Mg, and Cu atoms. Cu is involved in the early clustering process. • No GP II zones were clearly observed.
1	• GP I zones are still dominant.
24	• Large GP I zones were even stable during the course of T6 treatment, but the atomic fraction continuously decreases while η' fraction increases continuously.
0.5–4	• η' starts forming. Elongated small GP I zones transform to η'. The Zn/Mg ratio is 1.2–1.3. GP zones served as a heterogeneous nucleation sites for η' precipitates.
24	• η' reaches maximum number density or atomic fraction. GP I zones were also present but with lower number density.

the precipitation sequence as a function of time at the aging temperature of 121°C (250°F) using transmission electron microscopy and a 3-dimensional atom probe (3DAP). The details of their results are presented in Table 2.4.

2.3 EFFECT OF Ag

Like Cu, Ag also increases both the rate and level of hardening. The pioneering work of Polmear [16,17] has showed the positive effect of Ag addition on the strength of Al-Zn-Mg alloys. Both precipitate stability and the volume fraction of the strengthening precipitates were increased in the presence of Ag. In the case of Al-Zn-Mg-Ag and Al-Zn-Mg alloys, precipitation occurred after 3 min and 1 h, respectively, which clearly showed the effect of Ag on the hardening response [18]. Ag leads to the formation of Zn-Mg and Zn-Mg-Ag clusters at early stages of precipitation. There was no observation of Ag-Mg clusters. Ag also has strong affinity toward vacancies which promotes the formation Ag-vacancy clusters. Certainly, in the presence of Ag, the number density of GP zones increased. Unlike Al-Cu-Mg-Ag alloys, where Ag was found along the precipitate-matrix interface, in Al-Zn-Mg alloys Ag atoms were found in the precipitates [19]. Like Cu, Ag also increases the quench sensitivity and this may be due to the attraction between Ag atoms and vacancies, and faster aging response [20]. The combined additions of Cu and Ag resulted in more potent age hardening than individual additions as both the precipitation hardening kinetics and the precipitation number density are greatly enhanced.

2.4 EFFECT OF MICROALLOYING

In 7000 alloys, microalloying additions of Sc, Zr, Cr, Pr, and Ti are made for grain refinement during thermomechanical processing [21,22]. Generally, these intermetallic precipitates form during the conventional solution heat treatment process. The effectiveness of $L1_2$-type precipitates in increasing the tensile strength, providing high temperature stability due to low diffusivity, and Zener pinning of grain and subgrain boundaries are well proven. They also act as heterogeneous nucleating agents to form the strengthening precipitates (GP zones and η' precipitates). The major limitations in having a large density of these precipitates lie in their limited solubility in Al matrix. Therefore, the presence of low atomic fraction of these elements in Al alloys has to be efficiently utilized and the solutionizing treatments have to be optimized to maximize the effects. Cr addition has been replaced by Zr, as the latter reduces the quench sensitivity in thick Al alloy plates [20].

Zr is also most effective in inhibiting recrystallization. Zr-containing 7XXX alloys are known to have better strength–toughness relations as compared with Mn- or Cr-containing alloys. Both Sc and Zr form coherent $L1_2$-type precipitates. When Sc is present with Zr, the combination is more effective in inhibiting the recrystallization and increasing the alloy strength due to the presence of fine, coherent, and thermally stable $Al_3(Sc_{1-x}Zr_x)$ precipitates. Gao et al. [22] studied the influence of Ti on the age hardening behavior of Zr-containing Al-Zn-Mg-Cu alloy. In addition to the grain refinement, the Ti presence resulted in early hardening and a faster transformation from GP II zones to η' precipitates, and both of which resulted in higher peak hardness as compared to Zr-containing alloy.

2.5 EFFECT OF Li

The effect of Li addition in few of the Al-Zn-Mg-Cu alloys has also been investigated. Overall, Li addition to 7XXX series alloys is invariably detrimental for the reasons explained. The vacancy-binding energy of Li is very high, therefore Li atoms bind vacancies more effectively than other solute atoms (Zn, Mg, and Cu). As a result the formation of solute or vacancy-rich solute clusters of Zn, Mg, and Cu solutes are suppressed, thereby inhibiting the formation of metastable η'

strengthening precipitates. In these alloys, Al_3Li (δ') precipitates are the major strengthening phase and the remaining solutes form coarse T ((Al, $Zn)_{49}Mg_{32}$) and/or grain boundary precipitates. The strength maximization in these alloys can be obtained by predeformation and two-step aging treatment, where the defect structure in the grains provide effective nucleation sites for the $MgZn_2$-type precipitates along with fine δ' precipitates [23,24].

2.6 EFFECT OF PREDEFORMATION ON AGING

The presence of dislocations reduces the natural aging kinetics due to the annihilation of quenched-in vacancies at the dislocations. Moreover, the dislocations are favorable nucleation sites for the incoherent equilibrium η precipitates and the precipitate coarsening kinetics is also enhanced. The absorption of the solutes into the dislocations reduces the solute content that is available for the matrix precipitation reactions. The precipitate reaction in the matrix depends on the heating rate. Slow heating rate resulted in the formation of η' in the matrix along with the coarser η on the dislocations. In this case, the peak strength was slightly higher in the predeformed than the undeformed condition. On the other hand, when the heating rate was high, the formation kinetics of the η' was reduced because the GP zone formation tendency was reduced. Precipitate-free zone was observed near the dislocations along with coarser precipitates on dislocations and normal precipitate transformations far away from the dislocations. This resulted in reduced strengthening in the predeformed condition. A very large precipitate size distribution was observed in this case as opposed to the slow heating scenario where the distribution was narrower. In the overaging regime, the precipitates at dislocations got coarser at the expense of the smaller precipitate near them in both the slow and fast heating rate conditions. Therefore, the strength of the overaged alloy in predeformed condition is always lower than the undeformed condition [25].

2.7 SUMMARY

The summary of most of the discussion in this chapter is given in Fig. 2.1. The flow of the summary mostly follows the structure of the physical metallurgy discussion on the 7XXX alloys.

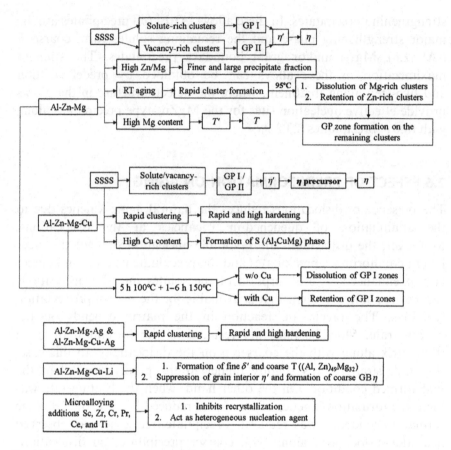

Figure 2.1 Summary of various reactions in Al-Zn-Mg-based alloys.

2.8 DIFFERENTIAL SCANNING CALORIMETRY OBSERVATIONS

For basic information on the differential scanning calorimetry (DSC) the reader is referred to Ref. [26]. In the case of 7449-T651 and 7150-T651 alloys, the DSC heat flow results were same even though 7449Al alloy contains relatively higher Zn content [27]. The peak position gets affected by the heat treatment condition. The peaks shifted to higher temperature in the overaged condition as compared to the peak-aged condition. In another study, the increase in Zn content from 9 to 11 wt.% in an Al-Zn-Mg-Cu alloy did not affect both the DSC peak positions and intensities [8]. The same observation can be made from another study as well and the results are shown in Table 2.5 [28]. T_p and T_e are the peak and end temperature of the η dissolution peaks, respectively. First, let us compare, alloy numbers 1 and 9, which differ

Table 2.5 Effect of Alloy Composition on the DSC Dissolution Peak of η' Precipitates [28]

Alloy No.	Zn	Mg	Cu	T_p (η') $(^{\circ}C)$	T_e (η') $(^{\circ}C)$
1	5.1	2.9	1.9	223	284
2	5.1	1.9	1.9	229	284
3	6.1	2.3	1.9	224	280
4	6.1	2.3	1.2	226	276
5	6.1	2.3	2.6	229	281
6	6.7	1.9	1.9	231	277
7	6.7	1.9	1.2	228	275
8	6.7	1.9	2.6	232	275
9	6.7	2.9	1.9	228	282

in Zn content and have same Mg and Cu content. There was no significant variation in the temperature. And the same conclusion can be made when only the Mg or Cu content varies in the alloys. Another major observation was the formation of S phase in alloys containing high Cu content. As discussed previously (in the effect of Cu subsection), even though microstructural characterization has concluded that Cu stabilizes the η' phase, the DSC observations could not capture that detail (Table 2.5).

REFERENCES

[1] < http://www.aluminum.org/sites/default/files/TEAL_1_OL_2015.pdf >.

[2] < http://www.aluminum.org/sites/default/files/Addendum_to_Teal_Sheets_DECEMBER_2014. pdf >.

[3] A. Dupasquier, R. Ferragut, M.M. Iglesias, M. Massazza, G. Riontino, P. Mengucci, et al., Hardening nanostructures in an AlZnMg alloy, Phil. Mag. 87 (2007) 3297–3323.

[4] L.K. Berg, J. Gjønnes, V. Hansen, X.Z. li, M. Knutson-wedel, G. Waterloo, et al., GP-zones in Al–Zn–Mg alloys and their role in artificial aging, Acta Mater. 49 (2001) 3443–3451.

[5] K. Stiller, P.J. Warren, V. Hansen, J. Angenete, J. Gjønnes, Investigation of precipitation in an Al–Zn–Mg alloy after two-step ageing treatment at 100° and 150°C, Mater. Sci. Eng. A270 (1999) 55–63.

[6] J.C. Werenskiold, A. Deschamps, Y. Bréchet, Characterization and modeling of precipitation kinetics in an Al–Zn–Mg alloy, Mater. Sci. Eng. A293 (2000) 267–274.

[7] S.P. Ringer, K. Hono, Microstructural evolution and age hardening in aluminium alloys: atom probe field-ion microscopy and transmission electron microscopy studies, Mat. Charact. 44 (2000) 101–131.

[8] Z. Chen, Y. Mo, Z. Nie, Effect of Zn content on the microstructure and properties of super-high strength Al-Zn-Mg-Cu alloys, Metall. Mater. Trans. A 44A (2013) 3910–3920.

[9] X.Z. Li, V. Hansen, J. Gjønnes, L.R. Wallenberg, HREM study and structure modeling of the η' phase, the hardening precipitates in commercial Al-Zn-Mg alloys, Acta Mater. 47 (1999) 2651–2659.

[10] G. Sha, A. Cerezo, Early-stage precipitation in Al−Zn−Mg−Cu alloy (7050), Acta Mater. 52 (2004) 4503–4516.

[11] J.Z. Liu, J.H. Chen, X.B. Yang, S. Ren, C.L. Wu, H.Y. Xu, et al., Revisiting the precipitation sequence in Al−Zn−Mg-based alloys by high-resolution transmission electron microscopy, Scr. Mater. 63 (2010) 1061–1064.

[12] T. Marlaud, A. Deschamps, F. Bley, W. Lefebvre, B. Baroux, Influence of alloy composition and heat treatment on precipitate composition in Al−Zn−Mg−Cu alloys, Acta Mater. 58 (2010) 248–260.

[13] X. Fang, Y. Du, M. Song, K. Li, C. Jiang, Effects of Cu content on the precipitation process of Al−Zn−Mg alloys, J. Mater. Sci. 47 (2012) 8174–8187.

[14] A. Dupasquier, R. Ferragut, M.M. Iglesias, F. Quasso, Vacancy-solute association in coherent nanostructures formed in a commercial Al-Zn-Mg-Cu alloy, Phys. Stat. Sol. 4 (2007) 3526–3529.

[15] T. Engdahl, V. Hansen, P.J. Warren, K. Stiller, Investigation of fine scale precipitates in Al−Zn−Mg alloys after various heat treatments, Mater. Sci. Eng. A327 (2002) 59–64.

[16] I.J. Polmear, A trace element effect in alloys based on the Aluminium-Zinc-Magnesium system, Nature (1960) 303–304.

[17] I.J. Polmear, K.R. Sargant, Enhanced age hardening in Aluminium-Magnesium alloys, Nature (1963) 669–670.

[18] S.K. Maloney, K. Hono, I.J. Polmear, S.P. Ringer, The effects of a trace addition of silver upon elevated temperature ageing of an Al-Zn-Mg alloy, Micron 32 (2001) 741–747.

[19] R. Ferragut, A. Dupasquier, M.M. Iglesias, C.E. Macchi, A. Somoza, I.J. Polmear, Vacancy-solute aggregates in Al-Zn-Mg-(Cu, Ag), Mater. Sci. Forum 519–521 (2006) 309–314.

[20] C. Nowill, Investigation of the quench and heating rate sensitivities of selected 7000 series aluminum alloys, Master's thesis, Worcester Polytechnic Institute, 2007.

[21] H.C. Fang, K.H. Chen, X. Chen, L.P. Huang, G.S. Peng, B.Y. Huang, Effect of Zr, Cr and Pr additions on microstructures and properties of ultra-high strength Al−Zn−Mg−Cu alloys, Mater. Sci. Eng. A 528 (2011) 7606–7615.

[22] T. Gao, Y. Zhang, X. Liu, Influence of trace Ti on the microstructure, age hardening behavior and mechanical properties of an Al−Zn−Mg−Cu−Zr alloy, Mater. Sci. Eng. A 598 (2014) 293–298.

[23] B.C. Wei, C.Q. Chen, Z. Huang, Y.G. Zhang, Aging behavior of Li containing Al−Zn−Mg−Cu alloys, Mater. Sci. Eng. A 280 (2000) 161–167.

[24] A. Sodergren, D.J. Lloyd, The influence of lithium on the ageing of a 7000 series alloy, Acta Metall. 36 (1988) 2107–2114.

[25] A. Deschamps, F. Livet, Y. Bréchet, Influence of predeformation on ageing in an Al-Zn-Mg alloy-I. Microstructure evolution and mechanical properties, Acta Mater. 47 (1999) 281–292.

[26] M.J. Starink, Analysis of aluminium based alloys by calorimetry: quantitative analysis of reactions and reaction kinetics, Int. Mater. Rev. 49 (2004) 191–226.

[27] N. Kamp, I. Sinclair, M.J. Starink, Toughness-strength relations in the overaged 7449 Al-based alloy, Metall. Mater. Trans. A 33A (2002) 1125–1136.

[28] X.M. Li, M.J. Starink, DSC study on phase transitions and their correlation with properties of overaged Al-Zn-Mg-Cu alloys, J. Mater. Eng. Per. 21 (2012) 977–984.

CHAPTER 3

Friction Stir Welding—Overview

3.1 INTRODUCTION

Friction stir welding (FSW) was invented at The Welding Institute, United Kingdom, in 1991 [1]. It is a solid-state welding technique, and Fig. 3.1 is an illustrative schematic explanation of the process. The example shows the plates to be welded in butt configuration. There are various other configurations possible such as T joint, lap joint, and fillet joint, and more information on joint configurations can be found in Ref. [2]. A nonconsumable rotating tool traverses along the butt line to make the weld by taking the material from the front of the tool and depositing at the back of the tool. The tool consists of shoulder and pin, and each has specific designs to aid in the material flow around the tool. For more information on FSW tools, the reader is referred to Ref. [2]. The primary functions of the FSW tool are threefold: (1) heating of the workpiece by the frictional heating between the tool and the workpiece and also by adiabatic heating, (2) movement of the flow-softened material around the tool to fill the gap behind the traversing tool, and (3) containment of the softened material under the tool shoulder.

3.2 TAXONOMY

A few important nomenclatures are defined and discussed below [3]:

Tool rotation rate: the rate at which the FSW tool rotates and is denoted as revolutions per min (rpm). This greatly influences the heat input in the workpiece and the material flow.

Tool traverse speed: the speed of tool travel along the weld joint and is expressed as inch per min (ipm). This also affects the thermal cycle and the material flow.

Tool tilt: the angle between the workpiece plane normal and spindle shaft and typically is between 0° and 3°.

Friction Stir Welding of High-Strength 7XXX Aluminum Alloys. DOI: http://dx.doi.org/10.1016/B978-0-12-809465-5.00003-9

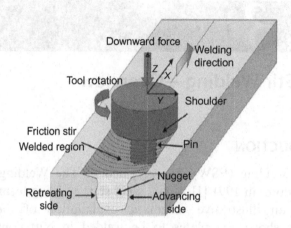

Figure 3.1 Schematic illustration of the FSW process [2]. Source: Reprinted with permission from Elsevier.

Plunge depth: it depends on the tool and is the depth of the FSW tool pin bottom from workpiece surface. It can be changed during the process based on the dynamic surface appearance of the weld.

Advancing side: the direction of the tool pin surface rotation and the tool traverse are the same.

Retreating side: the direction of the tool pin surface rotation and the tool traverse are the opposite. In Fig. 3.1, the FSW tool rotates in counterclockwise direction and traverses into the plane of the paper. Therefore, advancing and retreating sides of the weld are in the right and left, respectively.

Leading edge: the front side of the tool.

Trailing edge: the back side of the tool.

3.3 VARIOUS ZONES

FSW produces three distinct zones: the weld metal (nugget or stir zone (SZ)), the thermomechanically affected zone (TMAZ), and the heat-affected zone (HAZ). A description of the three zones plus the base metal is given below. The three FSW zones are identified based on different microstructural evolution and a macrograph displaying these regions is shown in Fig. 3.2.

Weld nugget (WN): the weld nugget is characterized by fine recrystallized equiaxed grains with variable (can be high or low) dislocation density. As it is in direct contact with the FSW tool, this region experiences both high temperature and severe plastic

Figure 3.2 A macrograph of the cross-section showing various regions in FSP 7075-T651 [2]. Source: Reprinted with permission from Elsevier.

deformation when compared to any other region. This region is also called SZ.

Thermomechanically affected zone (TMAZ): the TMAZ, as the name suggests, simultaneously experiences plastic deformation by the tool and thermal cycling, but to a considerably lesser extent, as compared to the weld nugget. As a result, there can be partial recovery and recrystallization depending on the strain and temperature experienced. This region lies adjacent to either side of the weld nugget. However, the width of TMAZ is higher on the retreating side as bulk of the material flows through that side.

Heat-affected zone (HAZ): the HAZ experiences only the welding thermal cycle (no plastic deformation). This region lies between the TMAZ and the base metal. Depending on the peak temperature experienced by this region and the initial temper of the base metal, the microstructural evolution can be quite different, and will affect the mechanical properties accordingly. The HAZ temperature continuously decreases from the start of the HAZ toward the unaffected base metal, hence the variation in mechanical properties.

Base metal: the base metal does not undergo a change in microstructure or mechanical properties. Though some part of this region might have experienced a thermal cycle, the temperature was not sufficient to induce any observable microstructural change.

3.4 MATERIAL FLOW

Material flow around the tool during the FSW process is highly complex. There are studies where either marker or tracer material or dissimilar welding was used to investigate the material flow. Colligan [4] investigated the material flow in 6061-T6 and 7075-T6 Al alloys using steel shot tracers filled in grooves that were parallel to the weld direction and were positioned at various distance and depths from the weld centerline. And then the positions of these steel shots after welding were analyzed using radiography. Two types of material flow behavior were observed: (1) chaotic deposition, where the material got stirred and extruded, was observed in the shoulder-controlled flow region, and (2) continuous deposition, where the material was simply extruded, was observed in pin-controlled flow region. The steel shots were either lifted up or pushed down slightly from their original z position. Guerra et al. [5] studied the material flow in 6061 using Cu foil marker. They observed three distinct zones: (1) rotational zones where the material rotates with the nib, (2) transition zone, and (3) deflection zone. They have also performed the lap welding of 2195-T6 (top plate) and 6061-T6 (bottom plate) Al alloys. Postwelding cross-section analysis has revealed a severe vertical movement of the two materials resulting in a banded microstructure. A short summary of the experimental evidences of the material flow is discussed in Ref. [3]. As can be noted, there is no unique observation of material flow and it may depend on the type of marker material or the tool geometry used.

The material around the pin also experiences complex strain and strain rate conditions. Strain and strain rate values calculated in A356 Al alloy were 3.5 and $85 \, s^{-1}$ at the shear boundary, where severe plastic deformation conditions are present [6]. Based on the model proposed by Long et al. [7] the strain at the retreating side is zero, increases continuously, and reaches maximum at the advancing side. In the friction stirring of Cu-Ag-Nb immiscible system, a huge variation in microstructural evolution and accumulated strain was observed in the stirred volume. The strain at the shear edge was around 6.5 and continuously decreased from the shear edge, reaching a zero value at the boundary between the deformed and undeformed volume [8]. The temperature distribution is also highly complicated and is presented in Chapter 4, Temperature Distribution.

3.5 DEFECTS IN FSW

Depending on the process parameter either a sound weld or a joint with defects will be obtained as shown in Fig. 3.3 [9]. In a hot weld, that is, high rpm and low ipm, the excessive material flow leads to flash formation, surface galling, and nugget collapse. Under cold

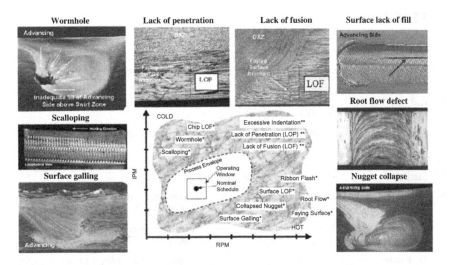

Figure 3.3 Various defects in FSW welds [9]. Source: Reprinted with permission from ASM International.

weld-processing conditions, that is, low rpm and high ipm, the lack of material flow will lead to surface lack of fill, wormhole, or lack of consolidation defects on the advancing side. Therefore, to produce a defect-free weld, optimum processing parameters, that is, optimum rpm and ipm, must be chosen.

3.6 KEY BENEFITS OF FSW

A detailed list of benefits of the FSW process is given in Ref. [2] and a few key points are highlighted below:

1. The high-strength, precipitation-hardened Al alloys which were deemed as nonweldable using fusion-welding techniques can be welded using FSW and result in excellent joint efficiencies (this point will be discussed in detail in Chapter 6, Mechanical Properties, under Section 6.4, Joint Efficiency).
2. Fine recrystallized microstructure with excellent mechanical properties of the weld.
3. Complete elimination of solidification-related microstructure and issues as it is a solid-state joining process.
4. Low distortion of the welded plates with good dimensional stability and repeatability.
5. No shielding gas is required and no harmful gas emissions.

REFERENCES

[1] W.M. Thomas, E.D. Nicholas, J.C. Needham, M.G. Murch, P. Templesmith, C.J. Dawes, G.B. Patent Application No. 9125978.8 (December 1991).

[2] R.S. Mishra, Z.Y. Ma, Friction stir welding and processing, Mater. Sci. Eng. R 50 (2005) 1–78.

[3] R.S. Mishra, P.S. De, N. Kumar, Friction Stir Welding and Processing, Springer International publishing, Switzerland, 2014.

[4] K. Colligan, Material flow behavior during friction stir welding of aluminum, Weld. J. 78 (1999) 229s–237s.

[5] M. Guerra, C. Schmidt, J.C. McClure, L.E. Murra, A.C. Nunes, Flow patterns during friction stir welding, Mater. Char. 49 (2003) 95–101.

[6] Z.W. Chen, S. Cui, Strain and strain rate during friction stir welding/processing of Al-7Si-0.3Mg alloy, IOP Conf. Ser. Mater. Sci. Eng. 4 (2009) 012026, 1–5.

[7] T. Long, W. Tang, A.P. Reynolds, Process response parameter relationships in aluminium alloy friction stir welds, Sci. Technol. Weld. Join. 12 (2007) 311–317.

[8] M. Komarasamy, R.S. Mishra, S. Mukherjee, M.L. Young, Friction stir-processed thermally stable immiscible nanostructured alloys, JOM 67 (2015) 2820–2827.

[9] W.J. Arbegast, A flow-partitioned deformation zone model for defect formation during friction stir welding, Scr. Mater. 58 (2008) 372–376.

CHAPTER 4

Temperature Distribution

4.1 INTRODUCTION

Among strain, strain rate, and temperature components during friction stir welding (FSW) process, the knowledge of temperature distribution in the stirred volume and around the processed zone is important as it controls the microstructural evolution. The major microstructural evolution are the change in grain size, grain size distribution, precipitate dissolution and reprecipitation and/or precipitate coarsening, and other phase transformations. FSW process parameters essentially control the microstructural evolution in various zones of the FSW weld. And not all precipitation-hardened Al alloys obey the same rules of the FSW process parameter−microstructural evolution relationship. For instance, in the case of Al-Mg-Sc-Zr alloys, the $Al_3(Sc, Zr)$ precipitates are extremely resistant to coarsening and there was no discernible heat-affected zone (HAZ) present [1]. Therefore, in these alloys, the FSW parameters are usually modified to change the grain size rather than to address the precipitate stability as is the case with 2XXX or 7XXX series precipitation-hardened alloys. In general, when all other parameters are kept the same, the peak temperature is mostly a function of tool rotation rate and the total time above a critical temperature is a function of the tool traverse speed.

4.2 EXPERIMENTAL OBSERVATIONS

There are quite a few studies on temperature measurement in the region just outside the welded region as it is relatively simple and avoids the tremendous complications arising due to the severe deformation in the stirred zone. Mahoney et al. [2] embedded several thermocouples close by to the weld nugget and measured the location-specific peak temperature, both as a function of the distance from the nugget and through the thickness, during the butt welding of the 7075-T651 alloy, is shown in Fig. 4.1. The arc on the right-hand side shows the edge of the FSW weld nugget. Near the edge of the

Friction Stir Welding of High-Strength 7XXX Aluminum Alloys. DOI: http://dx.doi.org/10.1016/B978-0-12-809465-5.00004-0

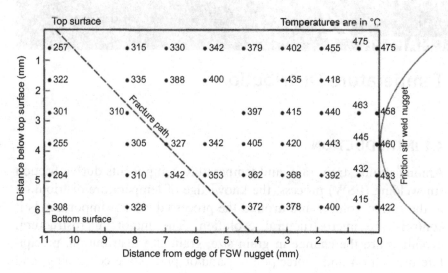

Figure 4.1 Temperature distribution adjacent to the FSW nugget of 7075-T651 [2]. Source: Reprinted with permission from Springer.

nugget, the peak temperatures varied from 415°C to 475°C. The highest peak temperature of 475°C was obtained near the top surface of the weld nugget edge. The peak temperature decreased as the distance from the weld edge increased. The lowest temperatures were recorded at a distance of ~11 mm from the weld edge and were between 257°C and 308°C. There were no information on the duration of the weld thermal cycle and the traverse speed was 5 ipm.

Colegrove and Shercliff [3] employed both the experimental and numerical methods to study the temperature distribution in 7075-T351 Al alloy. The location of the thermocouples is shown in Fig. 4.2A. Eight holes were drilled in each pair of the plates to be joined and one such plate is shown in Fig. 4.2A. Fig. 4.2B shows the variation in temperature in three locations as a function of tool rotation rate and welding speed. The temperature in the nugget, measured at location $y = 0$, $z = 13$, varied from 520°C to 550°C. As the temperature is really high and close to the solutionizing temperature, the temperature variation as a function of process parameters was not observed. At $y = 10$, peak temperature was high at higher rpm, and far from the weld no temperature variation as a function of tool rotation rate was observed. The computational fluid dynamics (CFD) results and the experimental observations followed a close trend, even though both the values did not match perfectly. They also showed that the advancing side peak temperature was slightly higher than the retreating side.

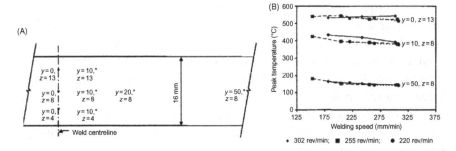

Figure 4.2 (A) Location of the thermocouples and (B) peak temperature variation in three locations as a function of the welding speed [3].

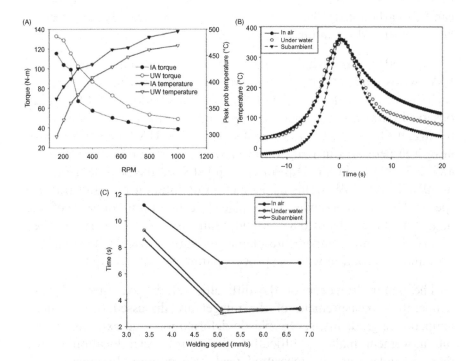

Figure 4.3 (A) Torque and peak temperature variation as a function of rotation rate under two conditions, (B) temperature–time at 8 ipm, and (C) time spent above 200°C in HAZ under all the three conditions [4]. Source: Reprinted with permission from Elsevier.

Upadhyay and Reynolds [4] studied the effect of various thermal boundary conditions on the nugget and the minimum HAZ location peak temperature. The nugget temperature was measured from the thermocouple welded to the probe and an external thermocouple was placed at the minimum HAZ hardness location (Fig. 4.3). The probe temperature and the torque in two thermal conditions as a function of the tool rotational rate are shown in Fig. 4.3A. The probe temperature increased and the torque decreased with the rpm. Moreover, underwater weld experienced lower probe temperatures at all the rotational rates. The subambient weld ($-25°C$) exhibited a sharp thermal cycle with minimum time at the peak temperature. On the other hand, the thermal cycle was the longest in air and it fell between these two extremes under water. Another point to note is that the peak temperature did not get affected by various thermal conditions. In the case of FSW of 7XXX Al alloys, both the peak temperature and the time above a critical temperature play a major role in precipitate transformations. The total time above 200°C temperature in various thermal conditions as a function of the welding speed is shown in Fig. 4.3C. The impact of external cooling can be clearly noted. The increased cooling rates have positive effect on both nugget and minimum HAZ hardness.

Fu et al. [5] also studied the temperature distribution in 7050 Al FSW welds made at 800 rpm and 4 ipm and under various conditions, such as in air, cold water ($\sim 8°C$), and hot water ($\sim 90°C$). They observed that the cold water cooling resulted in the lowest temperature in nugget (180–220°C) and no observation of HAZ minimum hardness. Temperatures in nugget and HAZ min were around 360–380°C and 275–300°C, in air. Hassan et al. [6] studied the peak temperature in 7010-T7651 FSW welds as a function of rotation rate and traverse speed. The thermocouples were placed just <1 mm outside of the nugget at the depths of 0.83, 3, and 5 mm from the top surface. They observed that the temperature at the top surface was higher than the root and its variation with the process parameter was as expected.

The major drawback or the difficulties of the experimental peak temperature measurement of the nugget are discussed below. Steep temperature gradients, severe plastic deformation, and extensive material movement make it difficult to know the exact location of the thermocouples and the measured peak temperature. In most of the cases, the thermocouples fail because of shear deformation. To circumvent the experimented complications, a lot of modeling and simulation efforts to investigate the temperature evolution in the stirred volume have been discussed in Section 4.3.

4.3 NUMERICAL OBSERVATIONS

An overview of the results obtained through various modeling and simulation studies will be discussed here. However, the details of various

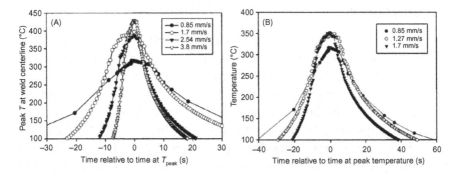

Figure 4.4 Temperature–time profiles (A) for welds with pitch 0.28 mm/rev and (B) for welds with 180 rpm [8].

models and methods that are used to obtain the temperature fields are not discussed. Frigaard et al. [7] employed a numerical 3-D heat flow model to study the temperature–time and temperature distribution results. They have also obtained the thermal cycle information from various locations near the weld using thermocouples. The computed thermal cycle did not match well with the measured values. Reynolds et al. [8] simulated the temperature–time profiles based on the experimentally measured torque information. The results are shown in Fig. 4.4. In Fig. 4.3A, the FSW condition with the fastest and the slowest welding speed exhibited the highest and lowest temperatures, respectively, at the same advance per revolution of 0.28 mm/s. Note that the rotation rate also increased along with the welding speed. Moreover, both the heating and cooling rates were exclusively a function of the welding speed. The temperature–time data for three welds that were made at different welding speed but at a constant rotation rate also showed the same trend.

Arbegast and Hartley [9] have developed an empirical model based on the tool rotation rate and traverse speed which have also been used widely to obtain the weld peak temperature. Ulysse [10] employed a 3-D viscoplastic model to study the temperature distribution in 7050-T7451 Al alloy and validated the model predictions with the measured thermal data. There was slight difference between the predicted and measured peak temperature. Nevertheless, the model predicted an increase and decrease in the peak temperature as a function of the increase in rotational rate and weld speed, respectively. Nandan et al. [11] have developed a 3-D material flow and heat transfer model to investigate the temperature, strain, and strain rate evolution during FSW. This approach was based on solving the conservation of mass, momentum, and energy equations with proper boundary conditions.

4.4 SUMMARY

Information about nugget peak temperature and temperature at HAZ (or the point of minimum hardness) from the available literature are summarized in Table 4.1. In the ambient conditions, the nugget temperature typically varies from 310°C to 480°C, except for the study by Colegrove et al. and both cold and hot water resulted in lower temperature as compared to the ambient conditions. In the case of HAZ, temperature at the minimum hardness point would greatly depend on the welding speed. The reported temperatures vary between 210°C and 370°C. But in majority of the studies, temperature above 300°C was commonly observed. The principal reason to investigate the temperature distribution in various weld zones is that the precipitate evolution is highly dependent on the local temperature. The added complexity in the case of 7XXX alloys, as was discussed in Chapter 2, Physical Metallurgy of 7XXX Alloys, the strengthening precipitates precipitate out at 121°C and the precipitate stability is extremely limited.

Table 4.1 Summary of Peak Temperatures Observed in Various Weld Zones					
Material	Process Parameter		Max Nugget T (°C)	T at HAZ HV_{min} (°C)	References
7075-T7651	5 ipm		>475	305−370	[2]
7075-T6	5 ipm		450−480	−	[12]
7075-T351 16 mm plate	255 rpm & 6, 8, 10, 12, 14 ipm		520−550	−	[3]
	302 rpm & 7.1, 9.5, 12 ipm				
	220 rpm & 7, 8.7, 10.4, 12.1 ipm				
7075-T451 6.35 mm	In air	250 rpm, 8 ipm	∼410	∼350	[4]
		650 rpm, 16 ipm	∼470	∼350	
		1000 rpm, 24 ipm	∼500	−	
	In underwater	250 rpm, 8 ipm	∼360	∼350	
		650 rpm, 16 ipm	∼450	∼350	
		1000 rpm, 24 ipm	∼470	−	
7050 Al	800 rpm, 4 ipm	In air	360−380	275−300	[5]
		In cold water	180−200	No HAZ	

(*Continued*)

Table 4.1 (Continued)

Material	Process Parameter		Max Nugget T (°C)	T at HAZ HV_{min} (°C)	References
7010-T7651	180 rpm, 3.7 ipm		Top = 401	–	[6]
			Center = 383		
			Root = 388		
	350 rpm, 3.7 ipm		Top = 476		
			Center = 437		
			Root = 424		
	280 rpm, 2.3 ipm		Top = 467		
			Center = 435		
			Root = 417		
7108-T79	1500 rpm	11.8 ipm	–	375	[7]
		18.9 ipm		285	
		28.3 ipm		250	
7050-T7451	180 rpm, 2 ipm		310	–	[8]
	360 rpm, 4 ipm		390		
	540 rpm, 6 ipm		390		
	810 rpm, 9 ipm		425		
	180 rpm	2 ipm	–	–	
		3 ipm	350		
		4 ipm	350		
7136-T76511	5 ipm	250 rpm	314	212	[13]
		350 rpm	415	276	
7050-T7451	700 rpm, 2.4 ipm		>430	260	[10]
	700 rpm, 4.5 ipm			210	
	700 rpm, 6.1 ipm			220	

REFERENCES

[1] S. Malopheyev, V. Kulitskiy, S. Mironov, D. Zhemchuzhnikova, R. Kaibyshev, Friction-stir welding of an Al–Mg–Sc–Zr alloy in as-fabricated and work-hardened conditions, Mater. Sci. Eng. A 600 (2014) 159–170.

[2] M.W. Mahoney, C.G. Rhodes, J.G. Flintoff, R.A. Spurling, W.H. Bingel, Properties of friction-stir-welded 7075 T651 aluminum, Metall. Mater. Trans. A 29A (1998) 1955–1964.

[3] P.A. Colegrove, H.R. Shercliff, Experimental and numerical analysis of aluminium alloy 7075-T7351 friction stir welds, Sci. Technol. Weld. Join. 8 (2003) 360–368.

[4] P. Upadhyay, A.P. Reynolds, Effects of thermal boundary conditions in friction stir welded AA7050-T7 sheets, Mater. Sci. Eng. A 527 (2010) 1537–1543.

[5] R.-D. Fu, Z.-Q. Sun, R.-C. Sun, Y. Li, H.-J. Liu, L. Liu, Improvement of weld temperature distribution and mechanical properties of 7050 aluminum alloy butt joints by submerged friction stir welding, Mater. Des. 32 (2011) 4825–4831.

[6] Kh.A.A. Hassan, P.B. Prangnell, A.F. Norman, D.A. Price, S.W. Williams, Effect of welding parameters on nugget zone microstructure and properties in high strength aluminium alloy friction stir welds, Sci. Technol. Weld. Join. 8 (2003) 257–268.

[7] Ø. Frigaard, Ø. Grong, O.T. Midling, A process model for friction stir welding of age hardening aluminum alloys, Metall. Mater. Trans. A 32A (2001) 1189–1200.

[8] A.P. Reynolds, W. Tang, Z. Khandkar, J.A. Khan, K. Lindner, Relationships between weld parameters, hardness distribution and temperature history in alloy 7050 friction stir welds, Sci. Technol. Weld. Join. 10 (2005) 190–199.

[9] W.J. Arbegast, P.J. Hartley, in: Proceedings of the Fifth International Conference on Trends inWelding Research, Pine Mountain, GA, June 1–5, 1998, p. 541.

[10] P. Ulysse, Three-dimensional modeling of the friction stir-welding process, Int. J. Mach. Tool Manu. 42 (2002) 1549–1557.

[11] R. Nandan, G.G. Roy, T.J. Lienert, T. Debroy, Three-dimensional heat and material flow during friction stir welding of mild steel, Acta Mater. 55 (2007) 883–895.

[12] C.G. Rhodes, M.W. Mahoney, W.H. Bingel, R.A. Spurling, C.C. Bampton, Effects of friction stir welding on microstructure of 7075 aluminum, Scr. Mater. 36 (1997) 69–75.

[13] C. Hamilton, S. Dymek, I. Kalemba, M. Blicharski, Friction stir welding of aluminium 7136-T76511 extrusions, Sci. Technol. Weld. Join. 13 (2008) 714–720.

CHAPTER 5

Microstructural Evolution

5.1 INTRODUCTION

Microstructural evolution links to the final properties because of the thermal cycle. In this chapter we start the discussion with the grain size evolution, continue to the dislocation density change, and finish with the most important of all, the precipitate evolution in various zones. A summary of the important changes in the microstructural features is given at the end for a quick reference.

5.2 EVOLUTION OF GRAIN SIZE

The grain size is a key basic feature of any microstructure and its importance for mechanical properties depends on the type of alloy. For ease of discussion, the information is presented with base metal as the baseline.

Base metal: The structure is partially recrystallized and exhibits pancake-shaped grains with subgrain structure. The basic structure comes from the rolling steps. The other shaping forms, like extrusion or forging, can impact the details but general characteristics are similar.

Nugget: The nugget or the stir zone is completely recrystallized and generally exhibits equiaxed grain structure. The grain size greatly depends on the friction stir welding (FSW) process parameters as it tracks the peak temperature. Finest grain sizes can be obtained at the lowest tool rotational rate and the grain size increases with the increase in rotation rate. For a particular rotational rate, the grain size increases with the decrease in the tool traverse speed. The nugget temperature varies from top to bottom and depending on the material thickness this can be significant; therefore a finer grain size is observed at the bottom of the nugget. The nugget undergoes dynamic recrystallization (DRX) as both the plastic strain and temperature experienced by the localized deforming volume are high. Su et al. [1] have

Friction Stir Welding of High-Strength 7XXX Aluminum Alloys. DOI: http://dx.doi.org/10.1016/B978-0-12-809465-5.00005-2

concluded that the continuous dynamic recrystallization (CDR) was the principal mechanism in refining the grain size. They have proposed the following mechanism to explain the recrystallization process. The dislocation density starts to increase from the instance the material comes in contact with the tool. After a critical dislocation density, subgrains start to form. With the continued deformation and further absorption of dislocations, the subgrains will grow and rotate, leading to the formation of recrystallized grains with high-angle grain boundaries. At this stage, there are two possibilities that can be expected. First, immediately after the completion of the recrystallization, if the processed material had exited the deformation zone, then the dislocation density in the grains would be lower. If not, further deformation in the processed zone will increase the dislocation density of the recrystallized grains. In fact, Su et al. [1] have observed grains with low dislocation density, grains that were partially recrystallized and also grains that had high dislocation density in the nugget region. Jata et al. [2] also observed that some nugget grains contained high dislocation density in 7050-T7451 Al alloy. On the other hand, Rhodes et al. [3] reported a low nugget dislocation density in 7075-T6 Al and later on, Rhodes et al. [4] performed rotating-tool plunge and extract technique combined with extreme surface cooling to have an insight into the nugget grain structure evolution. The grain size at the bottom of this stirred volume was around 25–50 nm, which is much finer than the subgrain size or the final nugget grain size. Based on this observation, they dismissed the recrystallization mechanism proposed by Su et al. [1] and concluded that the new recrystallized nanograins form in the nugget region. And these grains will undergo static growth during the weld cooling cycle after the tool has passed as was explained by Sato et al. [5] in 6063-T5 Al alloy. Furthermore, subsequent heating experiments of these nanocrystalline material had resulted in the grain sizes of 2–5 μm, which is equivalent to the ones observed in the nugget region. Subsequently, Su et al. [6] performed an experiment during the FSW of 7075-T6 Al alloy, where the tool was quickly removed at the weld end and the pinhole was quenched with the mixture of methanol and dry ice. They characterized the grain structure evolution near the exit hole and at various locations nearby. The region near tool contained grains of 100–400 nm in size, and the grain size gradually increased away from the tool. Based on these observations, they have proposed a new set of mechanisms. First, due to the complex state of deformation conditions near the tool, it is highly likely that discontinuous DRX lead to the formation of recrystallized nuclei and resulted in the formation of nanosized grains. Second, the growth of these grains or the introduction of further dislocations results in dynamic recovery. Third, the observation of the CDR. And finally after all these underlying mechanisms, the final FSW microstructure usually has grain size in the range of $1-5$ μm. As discussed earlier, the grains in the final microstructure will be at different deformation steps, which explain the variation in observed dislocation structure in the nugget grains. Furthermore, based on the observation of nanometer sized grains near the tool, Su et al. [7] conducted another experiment to freeze the microstructure of a normal weld by quickly quenching the plate behind the tool with the mixture of water, methanol, and dry ice. They have indeed obtained a bulk nanocrystalline material with the average grain size of 100 nm and with a large fraction of high-angle grain boundaries. Another unique observation in the microstructural evolution of precipitation-hardened Al alloys is the observation of abnormal grain growth (AGG). AGG was observed only at a certain combination of tool rotation rate and traverse speed. AGG in 7075 Al was explained using Humphrey's particle-pinning framework and the particle instability and the subsequent ineffectiveness in grain boundary pinning leads to AGG [8].

Another point to note is the nature of the grain boundary character distribution. The recrystallized nugget structure possesses a large fraction of high-angle boundary ($>80\%$) as opposed to other severe plastically deformed material where a large fraction of low-angle grain boundaries were reported. The existence of large fraction of high-angle grain boundaries is extremely important as it stabilizes the microstructure, both thermally and mechanically, and also leads to superplastic deformation behavior [9].

Thermomechanically affected zone: Su et al. [1] observed two distinct microstructural evolution in thermomechanically affected zone (TMAZ) and called them as TMAZ I and TMAZ II. The transmission electron microscopy (TEM) micrographs of the TMAZ I and TMAZ II regions are shown in Fig. 5.1D and E, respectively. TMAZ II, the region that is adjacent to the nugget, exhibited a recovered grain structure with equiaxed subgrains of $1-2$ μm, a low dislocation density, and low-angle subgrain boundaries. TMAZ I, the region that is close to the heat-affected zone (HAZ), contained coarse elongated grains

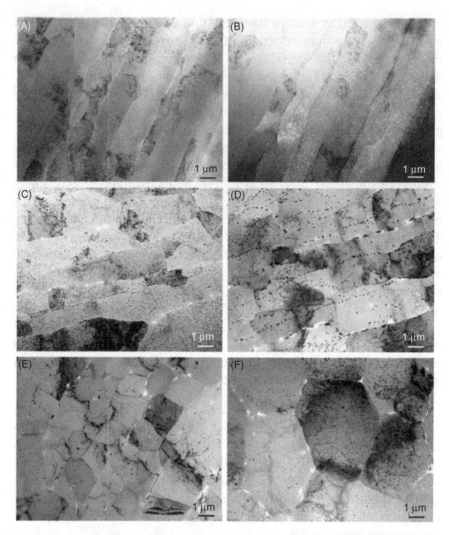

Figure 5.1 TEM micrographs of displaying the grain structure in different weld zones, (A,B) base metal, (C) HAZ, (D) TMAZ I, (E) TMAZ II, and (F) DXZ [1]. Source: Reprinted with permission from Elsevier.

with high dislocation density. The observation of the dislocation walls is indicative of the ongoing and incomplete dynamic recovery process in the TMAZ region. Rhodes et al. [3] observed no change in the grain or substructure in the TMAZ region when compared with the base metal. A low dislocation density was observed.

Heat-affected zone: Except for very slight coarsening, the HAZ grain structure resembles the base metal with low dislocation density [1].

5.3 PRECIPITATE EVOLUTION

In the case of precipitation-hardened Al alloys the major strengthening contribution comes from the precipitates. Therefore the evolution of precipitates in the localized region basically controls the strength. The precipitate evolution in various zones varies depending on the base metal temper and also the FSW process parameter. Furthermore, the peak temperature and the thermal cycle duration that a particular localized region experiences largely depend on the FSW process parameters. As discussed in Chapter 4, Temperature Distribution, the temperature variation in various weld zones is enormous. Also the precipitate evolution in the case of 7XXX alloys is highly sensitive to temperature, as noted in Chapter 2, Physical Metallurgy of 7XXX Alloys. The predicted η' solvus was around 355°C, and η solvus was around 438°C [10]. Therefore a detailed summary of the various transformation, such as complete/partial dissolution, coarsening, and the nucleation and growth of new precipitates, under various conditions is essential.

Nugget: Nugget experiences the highest temperature among all the zones, which is typically around ~ 450°C. A brief summary of the precipitate evolution in the nugget zone is given in Table 5.1. As can be noted from Table 5.1, there are two types of observations. First, depending on the peak temperature reached in the nugget region and its proximity to the solvus temperature, the precipitate structure might or might not dissolve completely. But, in the majority of the studies, a complete dissolution of the any precipitate (overaged η, Guinier–Preston (GP) zones) that is present.

Table 5.1 Literature Information Available on the Weld Nugget	
Observations	**References**
• Depending on the peak temperature, there can be partial or complete dissolution of the η precipitates • A negative response in the weld nugget to the postweld heat treatment (PWHT) was due to the formation of M phase, which is a nonstrengthening phase formed at high temperatures. This was observed in the welds that were made at very low welding speeds, that is, higher peak temperature and lower cooling rate	[11]
• Complete dissolution of the strengthening phase was observed in most of the studies • The dispersoids (Al₃Zr) did not dissolve during FSW. They might act as a nucleation sites for the strengthening precipitates during PWHT • The weld nugget was marked by the presence of GP zones, η' and coarse η after PWHT which results in considerable increase in the weld nugget strength	[1,2,12–14]

Table 5.2 Literature Information Available on FSW HAZ

Observations	References
Partial dissolution of the strengthening precipitates and no observation of coarsening in the HAZ	[12]
Mixture of unaltered η' and coarsened η due to the thermal exposure	[1]
Some fraction of precipitate dissolution was observed along with coarsened precipitates and widened precipitate-free zones. In one study GP zones were observed, which indicated the dissolution of precipitates during the heating cycle and reprecipitation during cooling cycle and/or via natural aging	[2,13,14]
Peak temperature in the HAZ region was between 330°C and 370°C. The hardness minimum was correlated with the welding speed. A lower welding speed tended to promote the formation of M phase. Limited dissolution and coarsening of the precipitates were also a characteristic of the HAZ	[11]

Heat-affected zone: A brief summary of the precipitate evolution in HAZ is given in Table 5.2.

A more detailed discussion based on quantitative observations and experiments is presented below. Dumont et al. [12] investigated the precipitate evolution in various FSW zones in 7449 Al alloy as a function of the initial base metal temper and the welding speed by employing TEM and small-angle X-ray scattering (SAXS). The base metal in T3 consisted mostly of GP zones, and T79 consisted mostly η and small quantities of η' as well. The difference in precipitate evolution can be noted from Fig. 5.2. SAXS maps of η precipitate volume fraction and size. None of the nugget precipitates had survived after the welding in both T3 and T79 base tempers. Therefore both the size and the volume fraction of the η are small. Based on TEM observations, GP zones is the major phase along with the stable η phase. The η phase and GP zones might have formed during the weld cooling cycle and natural aging. In the case of T79 HAZ, there was a reduction in volume fraction and also slight precipitate coarsening. And both η' and η precipitates were observed in TEM analysis. In the case of T79 TMAZ the precipitate volume fraction dropped further which signifies the precipitate dissolution. And the larger fraction of the remaining η precipitates decreases the tendency for GP zones formation. Therefore only coarsened η precipitates were observed. In the case of T3 condition and TMAZ, GP zones dissolve completely during the heating cycle. And, due to high temperature and the presence of large heterogeneous nucleation sites, the observed precipitates were of coarse η type. The size of η precipitates is larger in T3 condition as compared to T79 and this is due to the increased tendency for coarsening due to high supersaturation

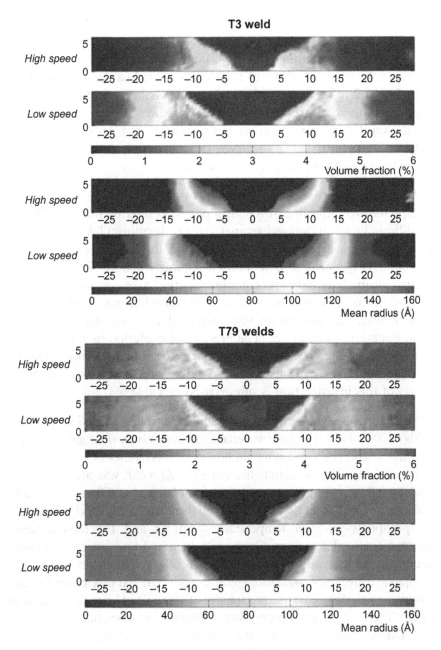

Figure 5.2 SAXS maps (volume fraction and size) of η precipitates in various zones for the T3 and T79 welds under low and high welding speeds [12]. Source: Reprinted with permission from Elsevier.

of solutes. Similar observation was noted in the case of HAZ as well. Though the pathway of the precipitate evolution was different among the T3 and T79 conditions, the final microstructures were more or less the same. Finally, the effect of the welding speed can also be deduced from Fig. 5.2. The zones in both the initial tempers become wider at lower welding speed as opposed to the narrower zone width at higher speed. Another prominent effect is the time at the peak temperature, which is lower in the case of higher welding speed. As a consequence, the coarsening of the precipitates is limited.

In another study on 7449 Al, Sullivan and Robson [10] studied the precipitate evolution in as-FSW and postweld heat treatment (PWHT) conditions in an underaged temper for age forming (TAF). The base material contained mostly η' precipitates in the size range of $1-7$ nm (3.5 nm average size) with very small volume fraction of η along with coarse grain boundary precipitates in the average size range of $10-300$ nm or 59 nm average size. And when the base material is T7 heat treated, the precipitate size increased to 5 nm with unaffected grain boundary precipitates. The precipitate evolution in various zones is presented in Fig. 5.3, and expectedly remarkable differences between the base metal and various zones were observed. The complete observation of this investigation is summarized in Table 5.3.

Su et al. [1] have investigated in detail the precipitate evolution in the FS welds of 7050-T651. The base material contained mostly η' $Mg(Zn, Cu, Al)_2$ with less η, and grain and subgrain boundary precipitates of η $MgZn_2$ and/ or $Mg_3Zn_3Al_2$. The precipitate-free zone (PFZ) width was around 25 nm. And the following precipitate evolution was observed in different zones and is shown in Fig. 5.4. In the nugget region, there was complete dissolution of the precipitates (nugget temperature $>450°C$) and the reprecipitation occurred on the dislocations. Grains with high and low dislocation density resulted in a high and low volume fraction of precipitates. In TMAZ the precipitate distribution was heterogeneous with bimodal precipitate size distribution consisting of coarse precipitates of 100 nm and fine precipitates of 10 nm. There was a clear observation of severe coarsening and dissolution in the TMAZ region due to the exposure to the high temperature ($350-400°C$). Moreover, the dislocation density was also high. Therefore the solutes tend to easily precipitate along the dislocations and subgrain boundaries during the cooling cycle (Fig. 5.5). Both GP zones and η phases were observed in TMAZ and DXZ regions. In HAZ the precipitates were mostly η with lower amount of η', distributed homogeneously in the

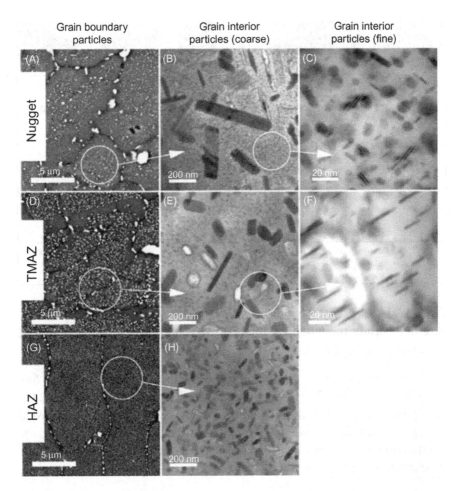

Figure 5.3 Postweld microstructural characterization of different weld zones, grain boundary precipitates in (A) nugget, (D) TMAZ, (G) HAZ; bimodal precipitates in (B) nugget, (E) TMAZ; fine strengthening precipitates in (C) nugget, (F) TMAZ; (H) coarsened HAZ precipitates [15]. Source: Reprinted with permission from Elsevier.

matrix, coarsened relative to the base material strengthening precipitates and also the PFZ width had increased by a factor of 5.

Another interesting study of precipitate evolutions in 7050-T7651 and 7075-T651 FSW welds was reported by Fuller et al. [14] where they observed the natural aging response of various weld zones. The summary of the observed precipitates are given in Table 5.4. Note that the natural aging was done for 73,300 h (~8.4 years). An interesting observation was the presence of GP zones in the HAZ region, which was completely absent invariably in all the remaining studies of

Zones	As-welded	PWHT (T7)
Table 5.3 Summarization of the Microstructural Evolution in Different Zones in 7449 TAF FSW Welds [10]		
Nugget	• Grain boundary (GB) precipitates increased from 10 to 300 nm in the base metal to 10–400 nm (average 144 nm) in the nugget • Grain interior precipitates increased from ∼3.5 nm in the base metal to ∼40 nm • Precipitates in the range of 3–4 nm also observed, possibly GP zones or η' precipitates • Overall, bimodal precipitate distribution was observed in the grain interior	• Transformation of fine GP zones and η' to coarse η • No change in the grain boundary precipitates
TMAZ	• GB precipitates increased from 59 nm in the base metal to 107 nm • Bimodal precipitate distribution with 50 nm coarser and 3.3 nm fine precipitates	• Transformation of fine GP zones and η' to coarse η • Slight coarsening of GB precipitates
HAZ	• GB precipitates increased from 59 nm in the base metal to 79 nm • Grain interior precipitates in size range of 9–60 nm	• Very slight coarsening of GB and coarse η grain interior precipitates

PWHT HAZ region. This clearly indicates the strong role of temperature in deciding the tendency for GP zones nucleation or coarsening of the existing η' or η precipitates.

A brief summary of the precipitate evolution is given in Fig. 5.6. As discussed earlier, irrespective of the base material temper, the precipitate size distribution is more or less same in difference weld zones and tempers. The positive effect of the natural aging on the strengthening precipitates can be clearly seen as compared to the precipitate evolution during artificial aging.

5.4 DIFFERENTIAL SCANNING CALORIMETRY

Differential scanning calorimetry (DSC) is widely used for the characterization of solid-state reactions, such as recrystallization, grain growth, shape memory effect, precipitation and dissolution, and other phase transformations; solid–liquid state reactions, such as incipient melting, melting, and solidification. For details on the experimental aspects and analysis, readers can refer to Ref. [15]. In the field of precipitation-hardened Al alloys, DSC technique has been extensively used in both quantitative and qualitative characterization of various aspects of the phase transformation. The DSC results contain endothermic and exothermic peaks. The endothermic peaks correspond to precipitate phase dissolution and the exothermic peaks correspond

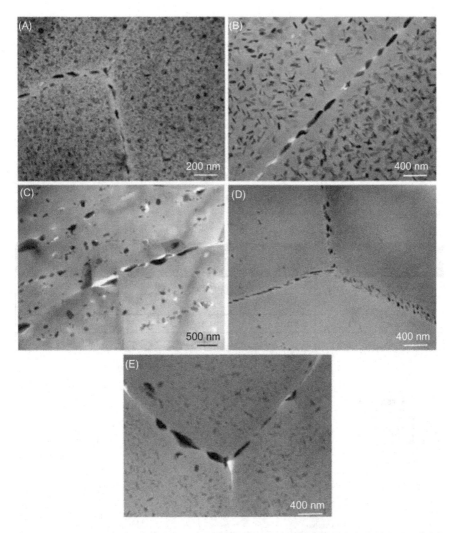

Figure 5.4 Precipitate structure in (A) base material, (B) HAZ, (C) TMAZ I, (D) TMAZ II, and (E) DXZ [1]. Source: Reprinted with permission from Elsevier.

to precipitate phase formation. The major output of DSC analysis are the following: (1) peak temperature (T_p) and end temperature (T_e) of various precipitation reactions that are mainly determined by the kinetics of the precipitate formation and dissolution, (2) area under the peak that is directly correlated to the volume fraction of the precipitates formed or dissolved, (3) activation energy for various precipitation reactions, and (4) the calculation of equivalent heat treatment time for the desired temperature based on the DSC data.

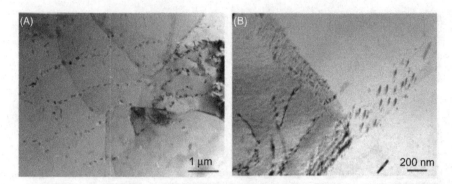

Figure 5.5 Precipitate structure in TMAZ (A) arranged like a deformed grain boundary and (B) preferential precipitation on subgrain boundaries [1]. Source: Reprinted with permission from Elsevier.

Table 5.4 Summary of Precipitate Evolution in HAZ and Nugget as a Function of Natural Aging Time [14]

	Approximate Natural Aging Time (h)		
	5–10	195	73,300
Heat-affected zone	GP(II) + η	GP(II) + η + ↑η'	↑GP(II) + η + ↑η'
Nugget	η'	↑GP(I) + η'	↑GP(I) + ↑η'

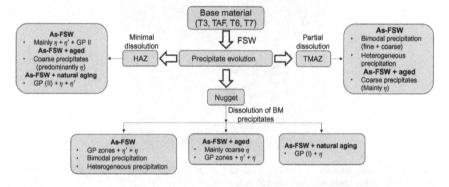

Figure 5.6 Overall summary of the precipitate evolution in different zones under various conditions.

The equivalent time approach is discussed briefly. Based on the DSC data, isothermal heat treatment equivalent times can be calculated for a given isothermal aging temperature using [15],

$$t_{eq} \cong 0.786 \frac{T_f}{\beta} \left(\frac{RT_f}{E} \right)^{0.95} \exp\left(-\frac{E}{RT_f} \right) \left[\exp\left(-\frac{E}{RT_{iso}} \right) \right]^{-1} \quad (5.1)$$

where T_f is the peak temperature, T_{iso} is the isothermal heating temperature, E is the activation energy for interested precipitation, R is the gas constant, and β is the heating rate. This approach in determining the time required to achieve the peak strength is additionally important in the case of FSW regions as the weld zones exhibit large variation in the supersaturation of solutes. The conventional T6 aging treatment to achieve the peak strength after FSW may not work as the microstructural condition across the zones vary a lot (underaged to overaged conditions) and might actually deteriorate the final strength of the weld. Both the peak and end temperatures depend on the heat treatment temper, solute supersaturation, and the DSC heating rate. Increase in heating rate shifts the DSC peaks to higher temperatures and thereby increasing both the peak and end temperatures of a particular precipitate reactions. Furthermore, the DSC profiles vary with the alloy temper. For instance, first peak in a solution-treated and solution-quenched sample would be the precipitation of the GP zones followed by the precipitation of other phases. An example of DSC thermogram for this condition is shown in Fig. 5.7 [16]. Most of the concepts discussed up to this point are noted in Fig. 5.7. First peak is the phase formation peak that is followed by various other reactions. Also, both the peak temperature and the area under the peak are also shown.

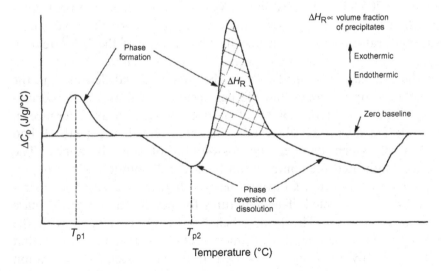

Figure 5.7 A typical DSC thermogram with baseline correction in the case of solution-treated and solution-quenched sample [16]. Source: Reprinted with permission from Springer.

In the case of naturally aged material the first peak would be the dissolution of the GP zones. Therefore DSC directly provides the signature of the type of precipitation reactions that can occur for a given sample condition during heating or cooling. Based on this information, the thermal history that a precipitation strengthened sample had undergone can be analyzed. Therefore this is an useful tool for analysis of precipitation in various FSW weld zones, because of the variation thermal cycle. The primary concern is to extract a representative sample of the localized region and also the repeatability of the results because a particular FSW run can have significant thermal gradient from the beginning of the weld to the end. Various DSC peaks and its temperature correlation are discussed here. GP I and GP II zones are usually observed between $50-140°C$ and $100-160°C$, respectively. The peaks that occur between $150°C$ and $300°C$ are correlated to the formation or the dissolution η' phase or the formation of η phase. η phase dissolution occurs between $300°C$ and $450°C$. At around $500°C$, the incipient melting of the S and/or T phase occurs and this depends on the chemical composition of the alloy.

Fuller et al. [14] studied the precipitate evolution using DSC as a function of natural aging time in both the weld nugget and HAZ zones of 7050-T7651 FSW joint. As discussed earlier, the peak temperature that these two zones undergo varies, hence the final solute supersaturation. Fig. 5.8 shows the DSC results of the HAZ region. The endothermic and exothermic peaks are denoted in Fig. 5.8. In the 5-h naturally aged sample the first peak labeled as 2 is the dissolution of GP zones and the next peak labeled as 3 is the formation of the η'. After 73,300 h of natural aging, a clear difference in the DSC peaks was observed. The area under the peak 2 had increased, which denotes the dissolution of the GP zones. The authors indicated the contribution from the dissolution of the η' precipitates in this peak. Conventionally a higher dissolution temperature for η' is expected. The signatures for peaks 2 and 3 in this case appear to overlap. The peak labeled as 4 was correlated with the formation of η precipitates during the DSC heating in samples that were naturally aged for a very long time. The precipitate evolution in the nugget zones is shown in Fig. 5.9. In the case of 7-h samples, only η formation was observed as there was no dissolution of GP zones that had precipitated out during this short time period.

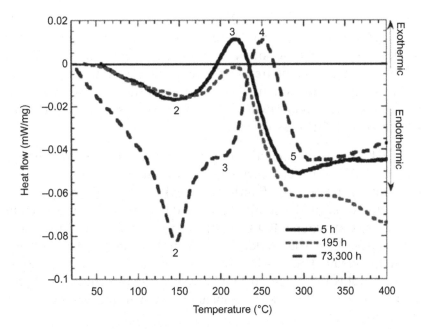

Figure 5.8 DSC peaks of the HAZ region of 7050-T7651 friction stir weld as a function of the natural aging time [14]. Source: Reprinted with permission from Elsevier.

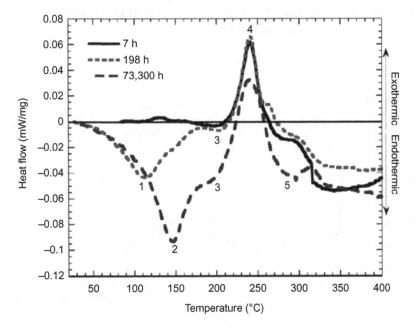

Figure 5.9 DSC peaks of the weld nugget region of 7050-T7651 friction stir weld as a function of the natural aging time [14]. Source: Reprinted with permission from Elsevier.

The endothermic peaks 1 and 2 correspond to the dissolution of the GP I and GP II zones that were observed in 198-h and 73,300-h samples, respectively. A smaller η formation peak was observed in 73,300-h sample as compared to the remaining conditions; this could be due to the reduction in the solute saturation in the matrix as the solutes were associated with a relatively stable GP II zone (with respect to GP I zone).

In 7042-T6 base metal the first peak observed was endothermic and at 165°C, which corresponds to the dissolution of the metastable η' phase [17]. Exothermic peak at 216°C corresponds to the precipitation of the stable η and/or the T phases. Another endothermic peak at around 400°C related to the dissolution of the η and/or T phases. Low rpm welds resulted in V-shaped welds and high rpm welds resulted in W-shaped welds. In the high weld energy or high rpm welds the postweld DSC revealed the presence of GP zone dissolution peak, which was completely absent in the low rpm welds (Fig. 5.10).

DSC is an excellent tool to capture the influence of the thermal history in various weld zones. The technique is fairly simple and when used with other microscopy techniques can result in a complete picture of precipitation in different weld zones.

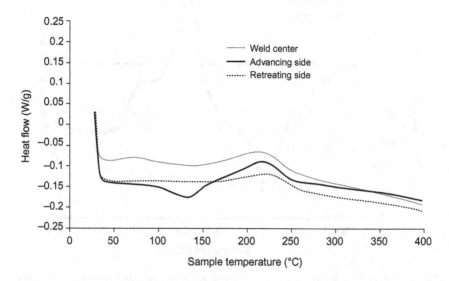

Figure 5.10 DSC peaks of weld center, advancing, and retreating sides of the 400-rpm weld [17].

5.5 SUMMARY

An overall brief summary of microstructural evolution is given in Table 5.5. The microstructural evolution is a strong function of various FSW process parameters that influence the thermal cycle. The recrystallized grain size is typically in the range of $1-10$ μm. By carefully controlling the process parameters and/or tool size, it is possible to obtain bulk nanocrystalline materials. The precipitate structure of the base material undergoes a complex evolution. For conditions resulting in peak temperature above precipitate solvus temperature, a complete dissolution of the

Table 5.5 Overall Summary of a Few Observations of Microstructural Evolution in 7XXX Alloys					
Material	Process Parameter	Zones	Grain Size	Precipitate Characteristics	Dislocation Density
7050-T651[1]	350 rpm, 0.6 ipm	Base metal (BM)	Millimeter-sized pancake-shaped grains	GP II, large and small fraction of fine η' and η, respectively	Relatively low
		HAZ	Millimeter-sized pancake-shaped grains	GP II, coarser η' and η	Same as BM
		TMAZI	Coarse elongated grains	Mainly η and GP I	High
		TMAZ II	Recovered, equiaxed subgrains of $1-2$ μm	Mainly η and GP I	Relatively low
		Nugget	Equiaxed grains of $1-4$ μm	Mainly η and GP I	Varies (low to high)
7075-T6 [3]	5 ipm	BM	Elongated grains	50−75 nm—Mg(Zn$_2$, AlCu) & Mg$_{32}$(Al,Zn)$_{49}$	Modest
				10−20 nm—η'	
		Nugget	$2-4$ μm	60−80 nm—Mg(Zn$_2$, AlCu) & Mg$_{32}$(Al,Zn)$_{49}$	Low
				GB precipitates— Mg$_{32}$(Al,Zn)$_{49}$	
				No 10 nm precipitates	
		TMAZ	Elongated grains	No change in 50−75 nm type	Low
				Coarsening of smaller precipitates	

(Continued)

Table 5.5 (Continued)

Material	Process Parameter	Zones	Grain Size	Precipitate Characteristics	Dislocation Density
7050-T7451 [2]	396 rpm, 4 ipm	Nugget–as-welded	$1-5\,\mu m$	No strengthening precipitates Al_3Zr	Varies (low to high)
		FSW + T6		Very fine precipitates	
		HAZ		Strengthening phases and PFZ are coarsened by 5	
7075-T6 [6]	350 rpm, 4.7 ipm	Nugget	$1-4\,\mu m$	–	High to low (differing)
7010-T7651 [18]	180 rpm, 2.3 ipm	Nugget	$1-3\,\mu m$	Volume fraction of the coarse second-phase particles decreased with increasing rotation rate	–
	450 rpm, 2.3 ipm		$5.5-6\,\mu m$		
	450 rpm, 7.7 ipm		$2.8-4.2\,\mu m$		
7075-T651 [13]	5 ipm	Nugget–as-welded	$3\,\mu m$	No strengthening precipitates	Low
				$MgZn_2$ of 100 nm in size	
		Nugget–as-welded + T6		Strengthening precipitates of 5 nm, PFZ of $\sim 50-100$ nm wide	
7075 Al [7]	1000 rpm, 5 ipm + quenching	Nugget	~ 100 nm	–	–

base material precipitates in the nugget region is followed by precipitation of various phases in varying size ranges depending on the cooling rate and PWHT conditions. Partial dissolution is normally observed in the TMAZ region followed by the observation of bimodal precipitate evolution during FSW and postweld aging. Finally, severe coarsening of the base material precipitates and a slight dissolution have been commonly observed in the HAZ region. As we start the discussion of mechanical properties in Chapter 6, Mechanical Properties, the evolution of precipitates in the HAZ has significant impact.

REFERENCES

[1] J.-Q. Su, T.W. Nelson, R. Mishra, M. Mahoney, Microstructural investigation of friction stir welded 7050-T651 aluminium, Acta Mater. 51 (2003) 713–729.

[2] K.V. Jata, K.K. Sankaran, J.J. Ruschau, Friction-stir welding effects on microstructure and fatigue of aluminum alloy 7050-T7451, Metall. Mater. Trans. A 31A (2000) 2181–2192.

[3] C.G. Rhodes, M.W. Mahoney, W.H. Bingel, R.A. Spurling, C.C. Bampton, Effects of friction stir welding on microstructure of 7075 aluminum, Scr. Mater. 36 (1997) 69–75.

[4] C.G. Rhodes, M.W. Mahoney, W.H. Bingel, M. Calabrese, Fine-grain evolution in friction-stir processed 7050 aluminum, Scr. Mater. 48 (2003) 1451–1455.

[5] Y.S. Sato, M. Urata, H. Kokawa, Parameters controlling microstructure and hardness during friction-stir welding of precipitation-hardenable aluminum alloy 6063, Metall. Mater. Trans. A 33A (2002) 625–635.

[6] J.-Q. Su, T.W. Nelson, C.J. Sterling, Microstructure evolution during FSW/FSP of high strength aluminum alloys, Mater. Sci. Eng. A 405 (2005) 277–286.

[7] J.-Q. Su, T.W. Nelson, C.J. Sterling, A new route to bulk nanocrystalline materials, J. Mater. Res. 18 (2003) 1757–1760.

[8] I. Charita, R.S. Mishra, Abnormal grain growth in friction stir processed alloys, Scr. Mater. 58 (2008) 367–371.

[9] Z.Y. Ma, R.S. Mishra, M.W. Mahoney, Superplastic deformation behaviour of friction stir processed 7075Al alloy, Acta Mater. 50 (2002) 4419–4430.

[10] A. Sullivan, J.D. Robson, Microstructural properties of friction stir welded and post-weld heat-treated 7449 aluminium alloy thick plate, Mater. Sci. Eng. A 478 (2008) 351–360.

[11] A.P. Reynolds, W. Tang, Z. Khandkar, J.A. Khan, K. Lindner, Relationships between weld parameters, hardness distribution and temperature history in alloy 7050 friction stir welds, Sci. Technol. Weld. Join. 10 (2005) 190–199.

[12] M. Dumont, A. Steuwer, A. Deschamps, M. Peel, P.J. Withers, Microstructure mapping in friction stir welds of 7449 aluminium alloy using SAXS, Acta Mater. 54 (2006) 4793–4801.

[13] M.W. Mahoney, C.G. Rhodes, J.G. Flintoff, R.A. Spurling, W.H. Bingel, Properties of friction-stir-welded 7075 T651 aluminum, Metall. Mater. Trans. A 29A (1998) 1955–1964.

[14] C.B. Fuller, M.W. Mahoney, M. Calabrese, L. Micona, Evolution of microstructure and mechanical properties in naturally aged 7050 and 7075 Al friction stir welds, Mater. Sci. Eng. A 527 (2010) 2233–2240.

[15] M.J. Starink, Analysis of aluminium based alloys by calorimetry: quantitative analysis of reactions and reaction kinetics, Int. Mater. Rev. 49 (2004) 191–226.

[16] J.L. Petty-Gails, R.D. Goolsby, Calorimetric evaluation of the effects of SiC concentration on precipitation processes in SiC particulate-reinforced 7091 aluminium, J. Mater. Sci. 24 (1989) 1439–1446.

[17] C. Hamilton, S. Dymek, O. Senkov, Characterisation of friction stir welded 7042-T6 extrusions through differential scanning calorimetry, Sci. Technol. Weld. Join. 17 (2012) 42–48.

[18] Kh.A.A. Hassan, P.B. Prangnell, A.F. Norman, D.A. Price, S.W. Williams, Effect of welding parameters on nugget zone microstructure and properties in high strength aluminium alloy friction stir welds, Sci. Technol. Weld. Join. 8 (2003) 257–268.

CHAPTER 6

Mechanical Properties

6.1 INTRODUCTION

Fig. 6.1 combines the concepts discussed in Chapters 4, Temperature Distribution, and 5, Microstructural Evolution, and leads to the current discussion on the mechanical properties. Fig. 6.1 shows a measured thermal cycle and the corresponding precipitate volume fraction evolution for a position of minimum heat-affected zone (HAZ) hardness. In the case of precipitation-hardened Al alloys the major strengthening contributors are the precipitates [1]. The precipitate evolution at a given location depends on the thermal history of that location. And the precipitate characteristics control the strength of the localized region. Finally both the localized and overall strength variation of the material control most of the overall mechanical properties.

6.2 HARDNESS AND TENSILE PROPERTIES

Mahoney et al. [2] studied the effect of friction stir process on the properties of 6.35-mm thick 7075-T651 Al alloy. The processed samples were aged at 121°C for 24 h (T6 treatment). They have carried out both longitudinal- and transverse-weld tensile tests in as-welded and T6-aged conditions. The longitudinal-weld samples were taken from the nugget region and consisted of a more uniform microstructure and the results are summarized in Table 6.1. The as-welded nugget showed a reduction in both yield and tensile strength as compared to the base metal, while the elongation was unaffected. The reduction in strength was mainly due to the dissolution of the strengthening precipitates. The postweld heat treatment had recovered most of the nugget strength, but there was a reduction in the ductility.

Table 6.2 shows the tensile test summary of the transverse-weld tests. The tensile properties were considerably lower as compared to the base metal and also the nugget region. The postweld aging had no effect on the yield strength as the HAZ strength had already reached the lowest

Friction Stir Welding of High-Strength 7XXX Aluminum Alloys. DOI: http://dx.doi.org/10.1016/B978-0-12-809465-5.00006-4

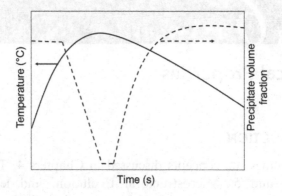

Figure 6.1 Weld thermal cycle measured at minimum HAZ hardness and the corresponding schematic of the precipitate evolution.

Table 6.1 Summary of the Longitudinal Nugget Tensile Properties of Friction Stir Welded 7075-T651 Al Alloy [2]

Condition	Yield Strength (MPa)	Tensile Strength (MPa)	Elongation (%)
Base metal	571	622	14.5
As-welded	365	525	15
Postweld T6 treatment	455	496	3.5

Table 6.2 Summary of the Transverse-Weld Tensile Properties of Friction Stir Welded 7075-T651 Al Alloy [2]

Condition	Yield Strength (MPa)	Tensile Strength (MPa)	Elongation (%)
Base metal	571	622	14.5
As-welded	312	468	7.5
Postweld T6 treatment	312	447	3.5

possible value due to the coarsened precipitates during welding; therefore a further heat treatment after welding did not alter the precipitate size. The reduction in elongation after postweld aging was due to the increased difference in strength between the lowest point in HAZ and the nearby nugget regions. The plastic deformation localization is severe in the postweld aging condition and resulted in low elongation values. Both the as-welded and aged transverse-weld tensile samples failed at the HAZ region. Contrary to the strength enhancement in the nugget region after aging treatment, the HAZ region did not respond to the aging treatment.

The strain distribution across the nugget and HAZ in the case of the as-welded sample is shown in Fig. 6.2 Note that the HAZ region

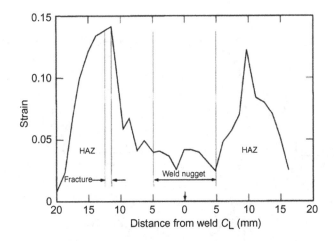

Figure 6.2 Local strain distribution in nugget and HAZ regions [2]. Source: Reprinted with permission from Springer.

exhibited the global minimum and the nugget region strength was higher. Due to high strength, the nugget region resisted the plastic deformation; therefore the low-strength region deformed extensively as can be seen in Fig. 6.2. The local strain level in HAZ and nugget had reached 12–14% and 2–5%, respectively. The strain localization in the HAZ region is the main reason for the observed low elongation values in the transverse-weld specimens.

Jata et al. [3] studied the mechanical property evolution of friction stir welded 7050-T7451 Al alloy. Fig. 6.3A shows the hardness data across the weld nugget at the top and root of the as-welded material. The through-thickness variation in the hardness profiles was due to the different thermal profiles at the top and bottom of the weld. The temperature difference leads to the variation in the extent of precipitate dissolution from top to the bottom of the weld. Minimum hardness was observed in the HAZ region. The effect of postweld aging on the top and root side of the weld is shown in Fig. 6.3B and C, respectively. The T6 aging treatment (121°C for 24 h) resulted in fine strengthening precipitates in the nugget and increased the nugget strength appreciably.

Furthermore, transverse-weld tensile tests were also performed in as-FSW, as-FSW + T6, and as-FSW + T7 (121°C for 8 h + 175°C for 8 h) conditions and summarized in Table 6.3. Due to the gradient in strength across the weld, the ductility of the weld had decreased. Also, both T6 and T7 aging treatments did not restore the tensile properties.

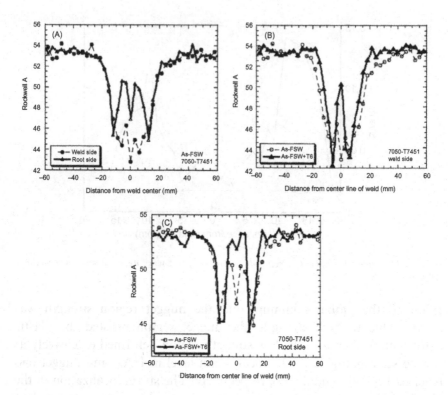

Figure 6.3 Hardness profiles across the weld at (A) weld top and root in as-FSW condition, (B) weld top in as-FSW and as-SFW + T6, and (C) weld root in as-FSW and as-FSW + T6 [4]. Source: Reprinted with permission from Springer.

Table 6.3 Summary of the Tensile Properties of the 7050-T7451 FSW Welds [3]

Condition	Yield Strength (MPa)	Tensile Strength (MPa)	Elongation (%)
Base metal	489	555	16.7
As-FSW	304	429	6
As-FSW + T6	291	417	3.8
As-FSW + T7	287	371	2.4

6.2.1 Effect of FSW Parameters

Reynolds et al. [4] systematically varied the tool rotation rate and traverse speed to study the hardness evolution as a function of the weld parameters. Fig. 6.4A shows the hardness variation across the weld as a function of the welding speed. Both the nugget and HAZ hardness were higher for the faster weld. Average as-welded nugget hardness as a function of the welding speed is shown in Fig. 6.4B. The nugget hardness at all the three advance per revolution increases

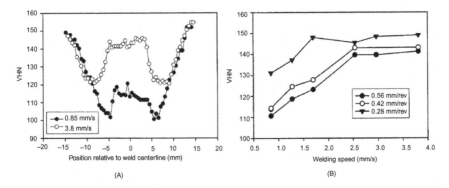

Figure 6.4 (A) As-welded hardness across the weld as a function of the weld speed for an advance per revolution of 0.42 mm/rev and (B) average nugget hardness as a function of welding speed for three advance per revolutions [4].

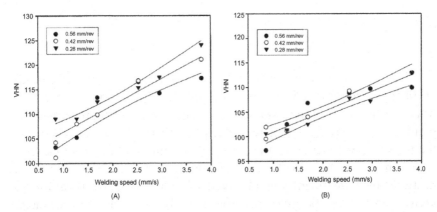

Figure 6.5 Variation in average minimum HAZ hardness as a function of the welding speed in (A) as-welded and (B) postweld heat-treated condition [4].

with the welding speed. This was related to the peak temperature experienced by the weld nugget and the corresponding dissolution of the stable precipitates.

The welding speed and the tool rotational rate also affect the HAZ minimum hardness. Fig. 6.5 shows the variation in the average minimum hardness as a function of the welding speed in (1) as-welded and (2) postweld heat-treated conditions. The minimum HAZ hardness increased with the welding speed. This is directly correlated to the extent of coarsening of the strengthening precipitates in HAZ. The extent of coarsening decreases with the reduction in time spent above a critical temperature and that is achieved by increase in the welding speed.

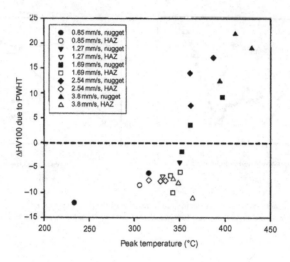

Figure 6.6 Change in hardness in nugget and HAZ in response to postweld heat treatment as a function of the peak temperature [4].

The HAZ exhibited a negative response to the postweld heat treatment at all the welding conditions, which indicates the absence of necessary or critical level of solute dissolved to precipitate out strengthening precipitates. The postweld aging response of both the HAZ minimum and the nugget as a function of the peak temperature and the welding speed is shown in Fig. 6.6. In the case of nugget, a positive response to the postweld heat treatment was observed when the peak temperature was above 355°C. The positive response was observed for the parameters with high tool rotation rate and fast welding speed.

Hassan et al. [5] studied the effect of welding parameters on the microstructure and mechanical properties of 7010-T7651 Al alloy. Fig. 6.7A shows the hardness profiles across the weld at the top, middle, and root for different tool rotation rates with a constant travel speed of 3.74 ipm. At the top, the tool rotation rate showed no influence on the nugget width. On the other hand, at the center and root, the nugget width increased with the tool rotation rate due to the increased heat input. Fig. 6.7B shows the effect of rotation rate and traverse speed on the nugget hardness. The lowest nugget hardness values were observed at the lowest tool rotation rates. A significant difference in hardness among the top, center, and root locations of nugget existed at the low tool rotational rates and slow traverse speeds. The hardness values for a given traverse speed increased, reached a

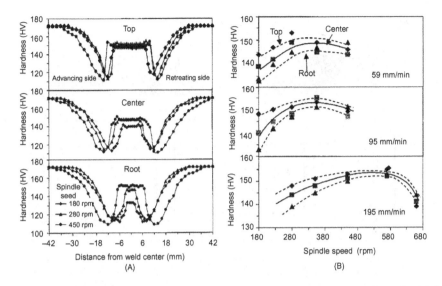

Figure 6.7 (A) Hardness profile across the weld as a function of the tool rotation rate or spindle speed and (B) effect of tool rotation rate on the hardness variation for the three tool traverse speeds [5].

plateau, and slightly decreased with the increase in the tool rotational rate. These trends indicate that maximizing the travel speed decreases the spread in hardness values from the top to the bottom of the nugget.

Vargas and Lathabai [6] also systematically studied the effect of tool rotation rate and traverse speed on the microstructure and mechanical properties of 7075-T6 Al alloy. Fig. 6.8A and B show the weld cross section and the corresponding hardness contour map, respectively. The effect of process parameters on the nugget hardness is shown in Fig. 6.8C–E. The effect of tool traverse speed on the nugget hardness (Fig. 6.8D) was more prominent than the tool rotation rate (Fig. 6.8C).

6.2.2 Effect of Thermal Boundary Conditions

The weld properties are generally controlled by the peak temperature and the total duration of the thermal cycle. As mentioned before, these two key characteristics are controlled by the tool rotation rate and the traverse speed. Further manipulation of the weld temperature and exposure time can be accomplished through the external cooling or heating. Upadhyay and Reynolds [7] made friction stir welds in 7050-T7451 that were submerged in water and at subambient temperature of $-25°C$

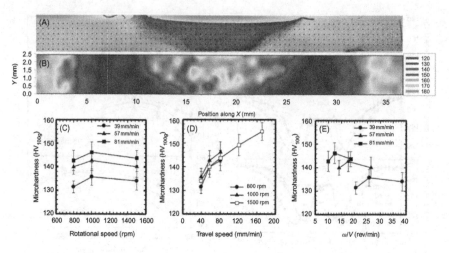

Figure 6.8 (A) Weld cross section, (B) hardness contour map, microhardness as a function of (C) tool rotation rate, (D) traverse speed, and (E) rev/min [6].

along with ambient conditions for comparison. The following postweld heat treatment was followed: 1 week of natural aging +121°C for 24 h. Fig. 6.9A shows the postweld heat-treated hardness across the weld that was made at 16 ipm and 800 rpm under various conditions. All the welds exhibit W-shape hardness profile. The HAZ hardness minima in subambient and under-water conditions were higher than the in-air weld. In the case of 200-rpm and 6-ipm weld, a clear distinction in hardness profiles between in-air and under-water conditions can be noted. The lower hardness and also a flat profile were most probably due to the precipitate coarsening as the peak temperature, which was near 350°C. In the regime of high peak temperature, both the under-water and subambient resulted in higher nugget hardness after postweld aging treatment as the quenching under these conditions were efficient as compared to the ambient condition. Therefore, beyond the intrinsic friction stir welding (FSW) process parameters, external cooling has proven to be an effective technique in obtaining a high joint efficiency weld. Similarly, Fu et al. [8] studied the submerged FSW of 7050 Al alloy in cold (8°C) and hot water (90°C). The effect of the heat extraction through cold or hot water can be seen in the hardness distribution across the weld (Fig. 6.10). The width of the low hardness zone is very large for welds performed in air. This is the basis for general thermal management during FSW, and a variety of schemes from cooling media to different thermal conductivity anvil can be employed to manipulate the overall thermal exposure.

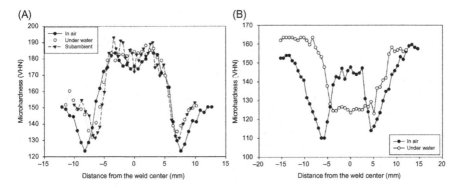

Figure 6.9 Postweld heat-treated hardness profile across the weld in (A) 800 rpm and 16 ipm and (B) 200 rpm and 6 ipm [7]. Source: Reprinted with permission from Elsevier.

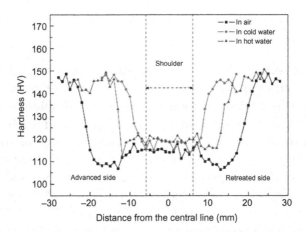

Figure 6.10 Hardness distribution across the weld under various conditions [7]. Source: Reprinted with permission from Elsevier.

6.2.3 Effect of Base Temper

There are a few studies that have concentrated on the effect of the initial base material temper on the weld properties. The tempers that are usually studied and compared are T7 [9], T6 [9–12], T4 [10], W [9,12], and O [10–12]. Note that the details on the temper designations are given in Chapter 2, Physical Metallurgy of 7XXX Alloys. Yan and Reynolds [9] studied the effect of base material temper (T7451, T62, and W) on the mechanical properties of a 6.4-mm thick 7050 Al alloy. This work was performed to identify the conditions under which a high joint efficiency weld can be obtained in 7XXX Al alloys. The authors pointed out that even though the postweld solution

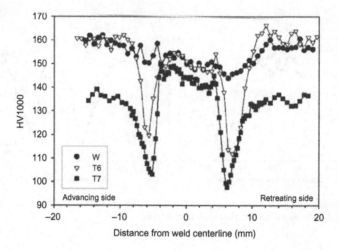

Figure 6.11 Hardness measurements across the weld mid-plane in W, T6, and T7 tempers after the postweld heat treatment [9].

treatment recovers the lost HAZ strength, it gives rise to the problem of abnormal grain growth in the nugget region. Moreover, subjecting the friction stir welded components to postweld solution treatment may not always be possible in many applications. Therefore, to circumvent all these issues, the authors proposed to study the role of base material temper on the joint strength. FSW was carried out in all three tempers using identical process parameters of 400 rpm and 11.2 ipm. All the three welds were immediately subjected to T7451 postweld heat treatment (121°C for 2 h + 163°C for 21 h). Both T6 and T7 tempers exhibited the commonly observed W-shaped profiles (Fig. 6.11). Nugget hardness of both the T6 and W tempers were the same, and the T7 nugget hardness was lower. Among all the three hardness profiles, W temper had resulted in a shallow a HAZ hardness minima with the difference being just 18 HV. The smaller difference between the minimum HAZ hardness and the nugget hardness, along with the width of the region of the minimum hardness, leads to better elongation and tensile strength. It is well known that the plastic deformation will initiate at the location of low strength, and in this case it is the HAZ region. Therefore, higher minimum HAZ hardness leads to higher transverse yield strength. The minimum HAZ hardness for the three welds are W = 144 HV, T6 = 112 HV, and T7 = 97 HV, respectively.

The effect of the hardness distribution on the tensile properties can be clearly distinguished in Fig. 6.12B. Fig. 6.12A displays the

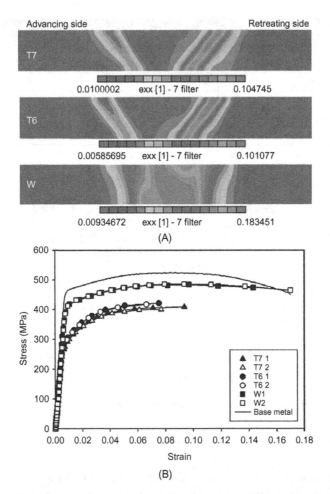

Figure 6.12 (A) Digital image correlation analysis of strain evolution across the weld during transverse-weld tensile test and (B) transverse-weld tensile test results of the base metals and the welds after postweld heat treatment along with the DIC stress–strain results [9].

digital image correlation (DIC) analysis of the strain mapping of all three transverse-weld tensile specimens just before the sample failure. The correlation between the location of the lowest hardness and the strain localization is remarkably clear. Tensile properties and the failure location of all the three transverse welds are summarized in Table 6.4. In the case of T6 and T7 tempers, based on hardness analysis, the retreating side HAZ was the weakest hence the strain localization. On the other hand, in the case the W temper, the difference in hardness was quite low and the width of the lowest hardness region was also large. Therefore, the entire nugget along with the HAZ region had plastically

Table 6.4 Summary of the Transverse-Weld Tensile Test Results Along With Failure Location [9]				
Condition	Yield Strength (MPa)	Tensile Strength (MPa)	Elongation (%)	Failure Location
Base metal	463	525	17	Base metal
T7	301	407	3.5	RS HAZ
T6	324	422	2.7	RS HAZ
W	401	486	8.8	Nugget

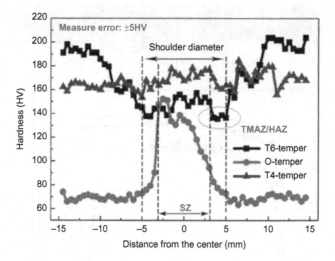

Figure 6.13 Across the weld hardness measurements of the friction stir welds in different initial base metal tempers [10]. Source: Reprinted with permission from Elsevier.

deformed as can be noted from the strain mapping (Fig. 6.12A). As a consequence, W temper is expected to have superior elongation as opposed to the T6 and T7 conditions and this can be observed in Table 6.4. Another point to note is that the transverse-weld yield strength values resembled the relative minimum hardness of all the three welds. A practical hindrance is to restrict the natural aging in 7XXX alloys which starts immediately after the solution treatment and to have the alloy in W temper to produce a high joint efficiency weld.

Chen et al. [10] studied both the microstructural and mechanical response of the Al-7B04 base metal in T4, T6, and O tempers after friction stir processing (FSP). The microhardness results are shown in Fig. 6.13, and tensile test results are summarized in Table 6.5. As expected, a W-shaped hardness profile was observed in T6 base metal condition after FSP. This is mainly due to the coarsening of the

Table 6.5 Tensile Properties of Base Metal and the Processed Material [10]				
Conditions	YS (MPa) (Transverse/ Longitudinal)	UTS (MPa) (Transverse/ Longitudinal)	Elongation (Transverse/ Longitudinal)	Failure Location (Transverse)
T6-BM	510	590	16.7	–
T4-BM	325	520	23.3	–
O-BM	110	220	20.0	–
T6-weld	(195/305)	(470/490)	(9.3/13.3)	HAZ (AS)
T4-weld	(315/320)	(485/495)	(13.0/12.0)	Nugget
O-weld	(115/145)	(210/390)	(12.5/10.7)	Base metal

strengthening precipitates in the HAZ region. The reduction in hardness across the processed region is marginally small in the case of T4 base metal temper condition. This can be attributed to the dissolution of the Guinier−Preston (GP) zones during the weld heating cycle and its subsequent formation during the weld cooling cycle. Major consequence is on the closeness between the base metal and longitudinal- and transverse-weld tensile strengths. And in the case of O temper, the weld thermal cycle leads to the precipitation of the strengthening phase in the nugget. Table 6.5 shows the tensile properties of the various base metal tempers and the corresponding processed material along with the transverse-weld failure location. The transverse-weld tensile samples failed in HAZ (AS), nugget, and base metal in T6, T4, and O tempers, respectively.

Furthermore, İpekoğlu et al. [11] also investigated the friction stir weldability of 7075 Al alloy in various starting tempers. 7075 Al alloy in O and T6 temper were selected for this study and FSW was carried out using different parameters. Irrespective of the FSW process parameters, hardness across the weld in both T6 and O temper conditions displayed the expected behavior. As expected, in O temper, hardness increase in the nugget was observed; on the other hand, nugget hardness decreased in T6 temper weld. Sharma et al. [12] also studied the temper effects (T6, W, and O) on the 7039 Al alloy welds and observed the same hardness response across the weld region as explained before. Therefore the alloy composition plays no significant role on the effects of the base metal temper on the microstructure and mechanical properties evolution. As we have already noted in Chapter 2, Physical Metallurgy of 7XXX Alloys, in the case of 7XXX alloys, the precipitate type and sequence does not get affected by the variation in the relative fraction of the alloying elements as it does in the 2XXX series Al alloys.

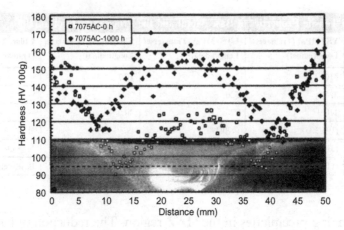

Figure 6.14 Microhardness of the actively cooled weld in as-welded and 1000 h naturally aged conditions [13].

6.2.4 Effect of Natural Aging

There are a few studies on the effect of natural aging on the mechanical properties of the friction stir welded 7XXX alloys. Nelson et al. [13] investigated the effect of 1000-h natural aging response on the mechanical properties of friction stir welded 7075-T351 Al alloy joined under various thermal conditions. They employed active heating and cooling which was imposed within 12 mm behind the tool, and passive heating and cooling which was applied on the anvil. Active cooling resulted in a fast cooling rate (25.6 K/s) with short time (21.9 s) above 200°C, which is the optimum condition to restrict coarsening of the existing strengthening precipitates or the reprecipitation of the dissolved precipitates. Fig. 6.14 shows the hardness mapping across the actively cooled weld in as-welded and naturally aged (1000 h) conditions. The nugget hardness after 1000 h of natural aging reached the base metal hardness in T3 condition, while the minimum HAZ hardness increased by more than 30 HV.

The tensile test results as a function of the natural aging time are shown in Fig. 6.15. It can be clearly seen that among the various thermal boundary conditions, the actively cooled condition exhibited comparatively better strength and elongation properties, followed by passively cooled condition. The tensile properties were inferior for the passively heated thermal condition as the cooling rate and the time above 200°C were the lowest (7.4 K/s) and longest (40 s), respectively. This was likely due to the severe coarsening of the strengthening precipitates as a result of such a thermal cycle.

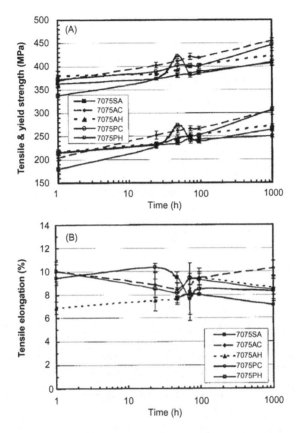

Figure 6.15 Tensile test results of the FSW 7075-T351 (A) yield and tensile strength and (B) elongation [13].

Fuller et al. [14] expanded the study conducted by Nelson et al. [13] to very long natural aging times and on 7050-T7651 and 7075-T651 Al alloys. Fig. 6.16A and B shows the Vickers microhardness profile for various natural aging time in 7050-T7651 and 7075-T651 welds, respectively. In both the alloys, the as-welded hardness profiles exhibited the commonly observed "W" shape. In both the welds, there was significant improvement in hardness after 216 h of natural aging. After 54,936 h of natural aging, hardness of the 7050 weld nugget (∼160 HV) was almost equal to the base metal (170 HV), while thermomechanically affected zone (TMAZ) and HAZ were slightly lower than the base metal hardness. Similar trend was observed in the case of the 7075-T651 Al weld. The hardness increase is most probably due to the formation and increase in volume fraction of GP (II) zones and η' precipitates.

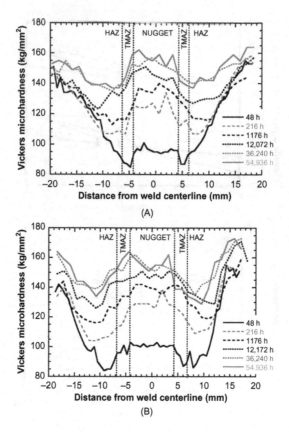

Figure 6.16 Hardness evolution across the weld as a function of natural aging times in (A) 7050-T7651 and (B) 7075-T651 Al alloys [14]. Source: Reprinted with permission from Elsevier.

Fig. 6.17 shows the variation in tensile properties as a function of the natural aging time in 7050-T7651 and 7075-T651 welds, respectively. The solid line and data point represent the transverse tensile properties, while the dotted line with open data points represents the longitudinal nugget properties. In both the welds, yield and tensile strengths increased with the increase in the natural aging time. In the case of transverse tensile tests the gage section consisted of zones of varying hardness values. In such cases the plastic deformation starts at the zone with lowest hardness. Depending on the strength difference among the neighboring zones and work hardening rates, the plastic deformation either spreads to other locations or gets localized. The localization results in faster failure, hence a lower observed elongation (which is measured on the basis of overall gage length) and tensile strength. On the contrary, in the case of the longitudinal

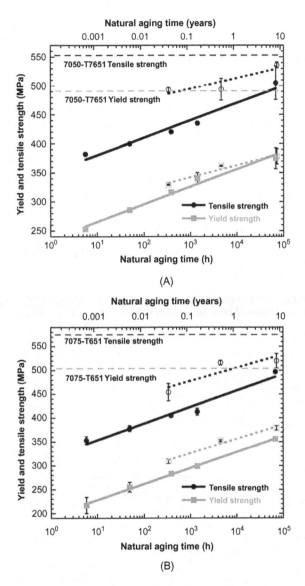

Figure 6.17 Tensile properties of (A) 7050-T7651 and (B) 7075-T651 Al alloys as a function of the natural aging time [14]. Source: Reprinted with permission from Elsevier.

sample, the gage contains a uniform microstructure, hence comparatively better tensile properties. The tensile properties of the longitudinal specimens, which were taken from the nugget region, were always better than the transverse-weld specimens.

The results of the tensile tests are also summarized in Table 6.6. HAZ strengths were interpreted from the transverse tensile test results as this is the only zone that would be deforming given the lowest hardness values.

Similar to Fuller et al. [14], Kalemba et al. [15] also investigated the long-term (6 years) natural aging behavior of friction stir welded 7136-T76511 Al alloy extrusions. 7136 Al alloy contains ~8 wt.% Zn. The base material hardness is around ~200 HV. The hardness increase after 6 years of natural aging in this study resembled the observations made by Fuller et al. [14] and is shown in Fig. 6.18. A significant increase in hardness was observed in both nugget and HAZ regions. Note that the minimum value has increased to the levels similar to the 7050 Al and 7075 Al alloys, even though it appears low on the plot because of higher base hardness.

Table 6.6 Summary of the Tensile Results as a Function of the Natural Aging Time [14]				
		Approximate Natural Aging Time (h)		
		5–10	195	73,300
Heat-affected zone	Yield strength (MPa)	255	305	365
	Tensile strength (MPa)	365	420	505
Nugget	Yield strength (MPa)	–	315	365
	Tensile strength (MPa)	–	480	535

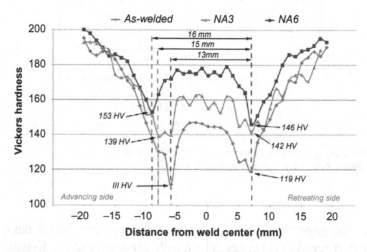

Figure 6.18 Mid-plane hardness across the weld in as-welded and naturally aged (3 and 6 years) specimens [15].
Source: Reprinted with permission from Elsevier.

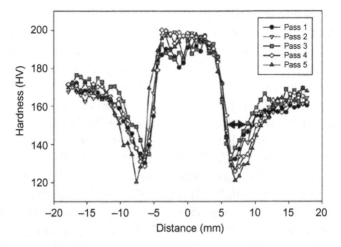

Figure 6.19 Transverse-weld hardness profiles of all the five passes [16]. Source: Reprinted with permission from Elsevier.

6.2.5 Multiple Pass FSW

Brown et al. [16] investigated the effect of multiple FSW passes on the properties in 7050-T7451 Al alloy. All the welds were done with the tool rotation rate of 540 rpm and the traverse speed of 16 ipm. Hardness distribution across the weld mid-plane for all the five weld passes after the postweld heat treatment (2 weeks of natural aging + 24 h at 121°C) is shown in Fig. 6.19. All the hardness profiles exhibited the typical W-shaped profile. The nugget hardness after postweld aging treatment was higher than the base metal hardness ($\sim$170 HV) in all the five passes. Hardness minima were observed in the HAZ region. Both the nugget grain size and hardness did not vary with the increase in the number of passes. On the other hand, in the case of HAZ, the effects of the multiple weld passes can be clearly observed. The HAZ minimum hardness on both the advancing and retreating sides decreased with the increasing number of passes. Furthermore, all the transverse-weld tensile properties decreased with the increasing number of welding passes. All the transverse-weld tensile samples from passes 1–4 fractured along the retreating side HAZ. Both the nugget tensile and base metal compressive residual stresses decreased with the increasing number of passes, although the measured peak temperature was same in all the five passes.

6.2.6 Effect of Material Thickness

In thick friction stir welds, as expected, a large variation in temperature between the top and root of the nugget is observed. High temperature near the weld top is due to the extensive heat generation at the shoulder and a lower temperature is observed at the bottom of the weld due to the backing plate heat extraction. The variation in temperature directly affects the microstructural evolution from top to bottom of the weld. Canaday et al. [17] investigated the through-thickness variation in properties of 32-mm thick 7050-T7451 friction stir welds. The following FSW parameters were used: tool rotation rate of 180 rpm, traverse speed of 2 ipm and 1° spindle tilt angle. The hardness profiles across the weld at three depths from the weld top in as-welded and postweld heat-treated conditions are shown in Fig. 6.20A and B, respectively. Both the conditions exhibited the usual W-shape hardness profile, typical of the precipitation strengthened Al alloys with high peak temperature. Postweld aging treatment resulted in a significant increase in hardness in the TMAZ region. Hardness increase after postweld treatment in nugget was relatively small. There was a slight difference in response between

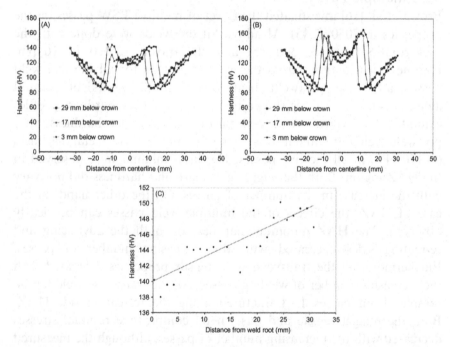

Figure 6.20 Transverse-weld hardness profile at three depth from the top surface in (A) as-welded, (B) postweld heat-treated conditions, and (C) variation in the weld centerline hardness as a function of the distance from the weld root in aged condition [17]. Source: Reprinted with permission from Elsevier.

the top and bottom of the nugget to the postweld heat treatment. In the case of HAZ, no change in hardness after postweld treatment was observed. Fig. 6.20C shows the variation in the nugget hardness as a function of distance from the weld root. The nugget hardness was low at the root and slightly increased toward the top of the weld. This is a clear indication of the variation in the microstructures between the top and bottom of the weld. Hence the peak temperature at the root is relatively less as compared to the top of the weld. The importance of this can gaged from the value of minimum hardness. In the thick material, the hardness drops to as low as 80 HV, whereas in thinner materials, the minimum hardness is in the range of 110−130 HV. Note that the low value of ∼80 HV is close to 50% of the parent material hardness of ∼160−170 HV.

Sullivan and Robson [18] performed half penetration FSW of 40-mm thick 7449 Al alloy in underaged temper for age forming (TAF) temper using Airbus Triflat tool. A subsequent T7 postweld heat treatment was applied. Transverse-weld hardness profiles at various depths from the top surface are shown in Fig. 6.21. There was no difference in the HAZ hardness from top to bottom of the weld. On the contrary, a huge variation in nugget hardness between the top and bottom of the weld can be clearly noted. Again, note the minimum hardness value of ∼90 HV. The lower hardness at the bottom of the weld was

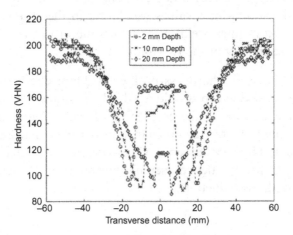

Figure 6.21 Transverse-weld profiles in as-welded condition at various depths from the top surface [18].
Source: Reprinted with permission from Elsevier.

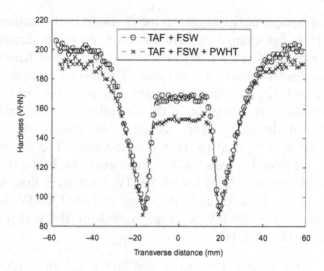

Figure 6.22 Transverse-weld hardness profiles in as-welded condition and T7 aged condition at 2 mm from the top surface [18]. Source: Reprinted with permission from Elsevier.

probably due to the low temperature experienced by this region that promoted severe precipitate coarsening with minimal dissolution. The nugget at the top and middle has experienced high temperature. This resulted in precipitate dissolution during the heating cycle and reprecipitation of the strengthening phases during the weld cooling cycle. A comparison of Figs. 6.21 and 6.20A shows that the variation in the through-thickness hardness values was larger in the 20-mm thick weld than in the 32-mm thick weld. Note that some of this is related to the difference in chemistry of the two alloys.

Fig. 6.22 shows the hardness profiles at 2 mm depth from the top surface in as-welded and postweld aged T7 condition. The reduction in nugget hardness is due to the coarsening of the strengthening precipitates. Surprisingly, there was no change in the HAZ hardness after T7 heat treatment. The base material which was in underaged temper had undergone the precipitate coarsening during the T7 treatment and resulted in hardness reduction.

6.3 FATIGUE AND DAMAGE TOLERANCE

Fatigue behavior is quite important for high-strength precipitation strengthened Al alloys. Structural components undergo varying stress

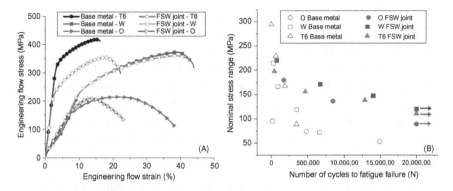

Figure 6.23 (A) Engineering stress—strain curves, and (B) S−N curves for base metals and FSW joints in various initial tempers [20]. Source: Reprinted with permission from Elsevier.

range during its life in service. Given the complexity and heterogeneity in microstructure, variation in properties across various weld zones, and also the presence of weld defects, investigation of the fatigue behavior of friction stir welded components is of importance.

There are two broadly employed fatigue investigation techniques. First is the total life approach, where the fatigue life is evaluated based on either the stress amplitude testing (S−N curve is the output) or the strain amplitude testing (plastic strain−N is the output). The number of cycles to failure under a specified stress or strain level is obtained. In this approach the total number of cycle includes both the crack initiation and propagation parts. The second method is defect-tolerance or the damage tolerance. The basic approach assumes that all the components contain preexisting flaws of some dimension. The total fatigue life in this case is the number of cycle necessary to propagate the crack into critical regime where the failure is catastrophic. For more information on fundamental mechanism on fatigue deformation, readers can refer to Ref. [19].

Sharma et al. [20] studied the effect of different base metal temper (T6, W, and O) on the fatigue behavior of friction stir welded 7039 Al alloy. Fig. 6.23A shows engineering stress—strain curves for the base metal and the welds in all the three tempers. As expected, T6 weld strength was lower than the base metal. A detailed account was given in subsection 6.2.3 (effect of base temper). With respect to the fatigue properties, the FSW joints exhibited superior properties as compared to their base metal counterparts. Furthermore, significant difference in the fatigue behavior of different tempers was also observed. The fatigue strength of the W and T6 tempers were slightly better than the O temper. The higher fatigue strength in W temper welds as compared to other tempers may be due to comparatively uniform and high-strength microstructure. The fatigue fracture location of the samples in different tempers matched with the lowest hardness point in the corresponding location across the weld hardness profiles. In T6, W, and O temper conditions, the weld specimens fractured along the TMAZ−HAZ interface, weld nugget, and the base metal, respectively.

The effect of kissing bond defect on the fatigue behavior of 7475-T7351 Al alloy was investigated by Kadlec et al. [21]. Kissing bond affected the static properties as the sample failed at low uniform elongation as compared to the one without a prominent kissing bond. DIC measurements of a flawless weld had strain localization at around the HAZ/TMAZ region, and the sample with the kissing bond exhibited

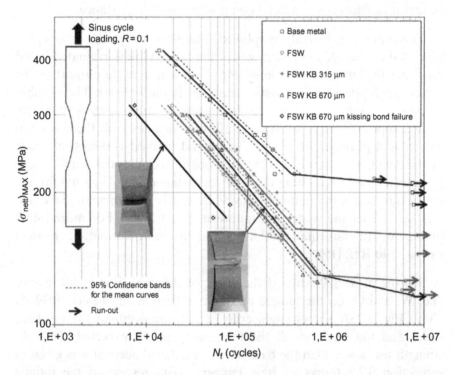

Figure 6.24 S−N curves for base metal, flawless FS welds, and welds with kissing bonds of 315 and 670 μm along with the inset showing two failure locations in 670-μm kissing bond [21]. Source: Reprinted with permission from Elsevier.

strain concentration at the kissing bond. Fig. 6.24 shows the S–N curves of the 7475-T7351 welds in flawless and flawed conditions. The base metal fatigue life values were superior to the flawless friction stir welds. Among the flawed welds that had kissing bond, the weld that had kissing bond of 670 μm presented a relatively large scatter in fatigue life data. The lowest fatigue life values were reported for the welds where the failure was initiated at the kissing bond.

A summary of fatigue failures in various conditions is given in Table 6.7. In most of the welds, as expected, failure occurred along the HAZ/TMAZ region. In the case of the 670-μm kissing bond, both the weakest region and the kissing bond defect initiated and aided in the sample failure.

Fig. 6.25 shows the fatigue crack growth (FCG) process. The left part shows the fracture surface of the kissing bond as well as the fatigue crack. The explanation is given in the right part. The fatigue

Table 6.7 Summary of the Fatigue Crack Initiation Locations Under Various Conditions [21]				
Material	Fatigue Crack Initiation			Kissing Bond Size (Mean Value)
	Base Metal	HAZ/TMAZ	Kissing Bond	
Base metal	X	–	–	–
FSW no defect		X	–	–
FSW small KB	X	X		315 μm
FSW large KB		X	X	670 μm

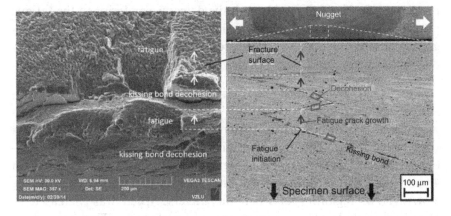

Figure 6.25 Observation and the explanation of the crack growth process [21]. Source: Reprinted with permission from Elsevier.

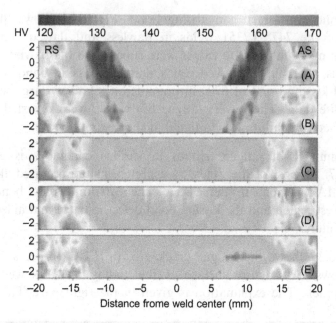

Figure 6.26 Microhardness contour maps of (A) 800 rpm and 4 ipm, (B) 800 rpm and 8 ipm, (C) 800 rpm and 16 ipm, (D), 1000 rpm and 16 ipm, and (E) 1200 rpm and 16 ipm [23]. Source: Reprinted with permission from Springer.

crack initiated in the middle of the zigzag shape and the kissing bond decohesion resulted in the crack propagation.

Various defects in the friction stir weld joint line affect the fatigue properties of the weld. Di et al. [22] studied the effect of zigzag curve defects on the fatigue properties of 7075-T6 FSW welds. The fatigue properties (S−N) of the welds with the defect was inferior to the fatigue properties of the flawless 7XXX welds. Most samples failed in the nugget region and the crack initiated from the weld root. This was mainly due to the zigzag curve defect observed at the bottom of the weld.

Feng et al. [23] studied the cyclic deformation behavior of friction stir welded 7075-T651 Al alloys. The following process parameters were used: 800- to 1200-rpm tool rotation rate and 4- to 16-ipm traverse speed. Fig. 6.26 shows the microhardness contour maps in various process conditions. Welds made using 800-rpm rotational rate with 4- and 8-ipm traverse speed exhibited region of low hardness in the TMAZ/HAZ region. And when the traverse speed was increased to 16 ipm, the low hardness zone decreased significantly. This is a critical observation as this can be directly linked to the fatigue crack initiation

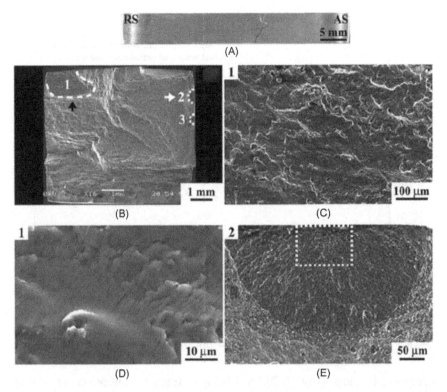

Figure 6.27 (A) Macroscopic view of the failed sample along with the crack, (B) overall view of the fracture surface with multiple crack initiation sites, (C) and (D) high magnification images of the location 1 which were close to the TMAZ region, and (E) location 2 initiation site which was along the nugget [23]. Source: Reprinted with permission from Springer.

behavior. At high strain amplitudes, the fatigue samples failed along the minimum hardness region in the case of 800-rpm and 4-ipm process condition. On the other hand, in the case of 800-rpm and 16-ipm welds, the fatigue failure occurred along the nugget region as there was no continuous low-strength TMAZ/HAZ region.

The stress amplitude increased with the number of fatigue cycle, which is an indication of the cyclic hardening. Higher cyclic hardening of the weld in comparison to that of the base metal was observed. Another observation is that the stress amplitude increased with strain amplitude. The stress amplitude of the friction stir welded joints was lower than the base metal and this is due to the lower yield strength of the welds. Fatigue cracks initiated from the specimen surface in both the slow and fast welding speed samples. Fig. 6.27 shows the fatigue fracture surfaces of the 800-rpm and 16-ipm weld along with the

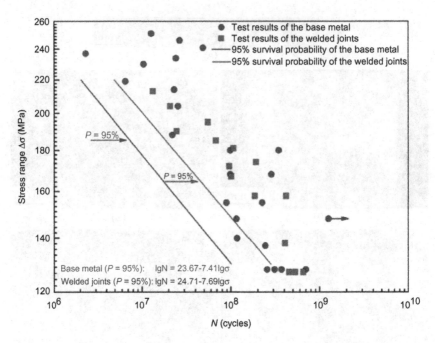

Figure 6.28 S−N curves for the base metal and the FSW joint [24]. Source: Reprinted with permission from Elsevier.

macroscopic view (Fig. 6.27A). Multiple fracture initiation sites were observed as circled in Fig. 6.27B. Location 1 was in TMAZ and the crack initiated at the top surface near TMAZ are shown in Fig. 6.27C and D. Fig. 6.27E shows the crack initiation in the nugget region. The crack initiation was associated with intergranular cracking. Fatigue failures initiation at the surface roughness, flaws, or on flash is quite common.

Deng et al. [24] studied the effect of the microstructural and strength heterogeneity of the friction stir weld on the very high cycle fatigue behavior of FSW 7050-T7451 using ultrasonic fatigue tests. Fig. 6.28 shows the stress amplitude−number of cycles to failure curve for the base metal and the FSW joint. The fatigue behavior of the FSW joint was the same as the base material in the ultrahigh cycle fatigue regime. Furthermore the scatter in the fatigue behavior of the joint is smaller than the base metal. Fig. 6.29 shows the fatigue fracture positions of the FSW joint. The failure, in most cases, occurred along the weakest link in the precipitation-hardened FSW joint, TMAZ/HAZ locations. Advancing side had more number of failures than the retreating side. This is due to the relatively lower TMAZ/HAZ hardness at the advancing side. Feng et al. [23] also observed the fatigue failure along the TMAZ/HAZ boundary in the 7075-T651 FSW joints.

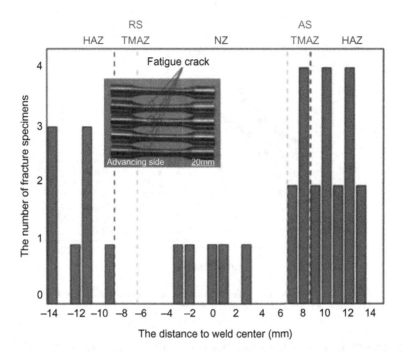

Figure 6.29 Survey of fatigue failure locations in FSW joints [24]. Source: Reprinted with permission from Elsevier.

FCG behavior of the nugget and HAZ regions in as-FSW + T6 condition were investigated and compared with the 7050-T7451 Al base metal (Fig. 6.30) [3]. At higher stress ratio, $R = 0.7$, the FCG resistance behavior in the near-threshold regime of the base metal, nugget and the HAZ region were similar. On the other hand, at low stress ratio ($R = 0.33$), the FCG resistance of the nugget along the weld centerline was lower than the base material. Interestingly, FCG resistance of HAZ region was higher than the base metal. Fatigue crack closure near the threshold regime was completely absent in the case of base metal and the nugget, and was significant in the case of the HAZ region.

In general the following crack growth behavior in the base metal was observed. Precipitate size and coherency play a critical role as these determine the slip reversibility characteristics and control the accumulated damage. In the case of underaged precipitates, the dislocations shear the coherent precipitates that result in a higher slip reversibility and lower accumulated damage. On the other hand, in overaged precipitates, the dislocations loop around the coarser, incoherent precipitates,

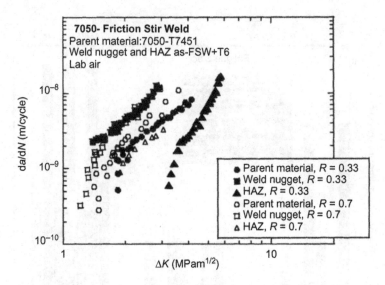

Figure 6.30 Fatigue crack growth behavior of nugget, HAZ and parent material, tested at R = 0.33 *and* R = 0.7 *stress ratios [3].* Source: Reprinted with permission from Springer.

and cannot reach back to the crack tip during the stress reversal. This results in the damage accumulation that leads to faster crack growth. Therefore alloys with underaged precipitates undergo severe crack closure and a better FCG resistance as compared to the alloys with overaged precipitates. Accordingly the FCG resistance of the nugget in T6 condition was expected to be better than the HAZ region, which was not the case. Hence the observed behavior cannot be explained based on the microstructural state alone. It is known that the residual stresses in friction stir welds are lower than the fusion welds, but even a low residual stress state in various weld zones might affect the FCG behavior. The residual stress were measured on the eccentrically loaded single-edge-tension (ESE(T)) FCG specimen as shown in Fig. 6.31A. The results of the longitudinal and tensile residual stress components are shown in Fig. 6.31B and C, respectively. The transverse residual stress was negligible as compared to the longitudinal residual stress component. Both HAZ and weld centerline (nugget) exhibited a compressive residual stress along the longitudinal direction. Therefore, the residual stress in this case should enhance the FCG resistance in both HAZ and nugget by closing the crack. As the expected behavior based on the residual stress argument was obtained in the HAZ region, the nugget FCG behavior cannot still be explained. Both the positive factors, coherent precipitates and compressive residual stress, did not control the

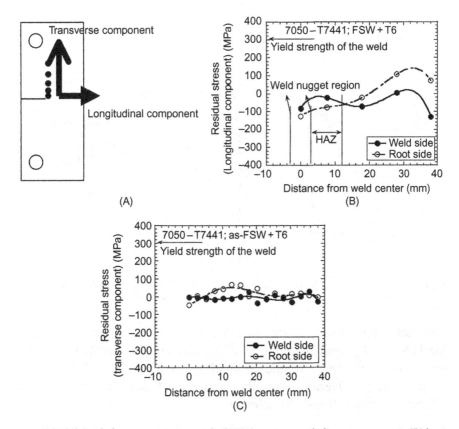

Figure 6.31 (A) Residual stress measurements on the ESE(T) specimen, residual stress components in (B) longitudinal, and (C) transverse directions [3]. Source: Reprinted with permission from Springer.

FCG behavior in the nugget region (weld centerline). The fatigue fracture surface revealed the presence of extremely fine fracture facets and also intergranular cracking. To conclude, the FCG behavior in the weld nugget region was dominated by the fine grain size and a compressive residual stress state in HAZ region.

John et al. [25] studied the residual stress effects on the near-threshold FCG in friction stir welded 7050-T7451 Al alloy. The FCG behavior was studied as a function of the microstructure, specimen geometry, stress ratio, and residual stress. At low R (stress ratio), the crack growth behavior was geometry dependent and at high R, the geometry dependence is eliminated, Fig. 6.32. In HAZ, the FCG resistance should decrease due to the overaged precipitates. But the compressive residual stresses play major role in determining the FCG.

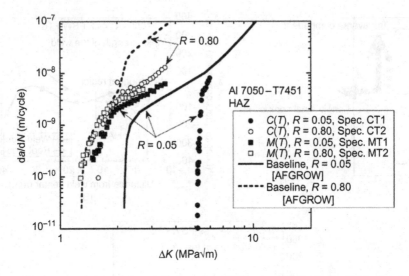

Figure 6.32 Fatigue crack growth behavior of base metal and HAZ region [25]. Source: Reprinted with permission from Elsevier.

The damage tolerance of the nugget region was relatively lower than the HAZ and parent material. The degradation was attributed to the fine grains. Furthermore the residual stress also affected the FCG through K_{max}. Overall, even the presence of low residual stresses affects the fatigue threshold behavior.

Uematsu et al. [26] investigated the fatigue behavior of 7075-T6 Al welds in air and 3% NaCl solution. The fatigue life of the base metal and the FSW welds were similar in air at room temperature. Under salt water, both the base metal and FSW joint fatigue strengths decreased as compared to the in-air testing condition. Fatigue cracks initiated along the TMAZ/HAZ region and deep corrosion pits were observed at the initiation sites.

6.3.1 Effect of Laser and Shot Peening

The laser and shot peening effects in tackling the weld residual tensile stresses by incorporating the compressive stresses have been reported. The residual stress in a FSW joint is lower as compared to the fusion welds, nonetheless, even a low residual stress impacts the fatigue behavior. A detailed analysis of the impact of residual stresses on the FSW joints has been reported [27]. Hatamleh et al. [28] studied the laser and shot peening effects on the FCG behavior of 7075-T7351 Al FSW joints and the results are shown in Fig. 6.33. The FCG behavior

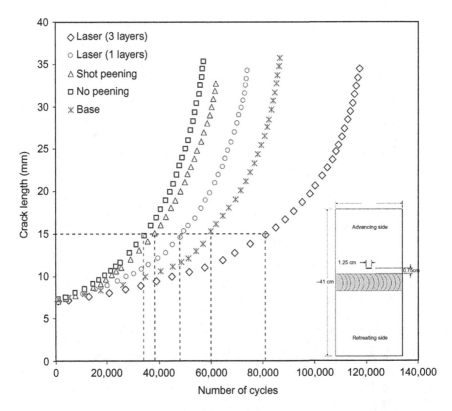

Figure 6.33 Variation in crack length with number of fatigue cycles for various samples [28]. Source: Reprinted with permission from Elsevier.

of the as-welded, shot peened, and laser peened with one layer were lower than the base metal. The sample processed with three layers of laser peening exhibited significant increase in the FCG resistance. Better fatigue properties in the three layers of laser peening specimen were attributed to the introduction of higher and deeper compressive stresses as compared to shot peening. Furthermore, the FCG rate in the three-layer laser peened was significantly lower than the remaining material conditions. The slower FCG rate is indicated by fine striation spacing on the fatigue fracture surface of the three-layer laser peening.

In another study, Hatamleh et al. [29] studied peening effects on the FCG behavior of a 12.5-mm thick 7075-T7651 Al FSW joints. The surface longitudinal and transverse residual stresses measured by X-ray diffraction method are shown in Fig. 6.34A and B, respectively. Furthermore the peening has eliminated the tensile residual stresses and introduced compressive stresses, in both conditions. The crack

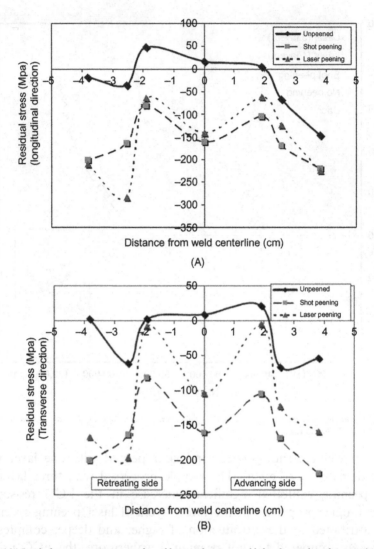

Figure 6.34 Residual stress mapping across the weld top surface in as-welded and peened conditions in (A) longitudinal and (B) transverse orientations [29]. Source: Reprinted with permission from Springer.

length variation as a function of the number of the fatigue cycle for the stress ratio of 0.1 is shown in Fig. 6.35. Both the as-welded and shot peened joints exhibited faster crack growth as compared to the base metal. On the other hand, laser shot peened material displayed slower crack growth and showed greater FCG resistance. In the case of a higher stress ratio ($R = 0.7$), both the laser and shot peened welds exhibited superior FCG resistance as compared to the base metal.

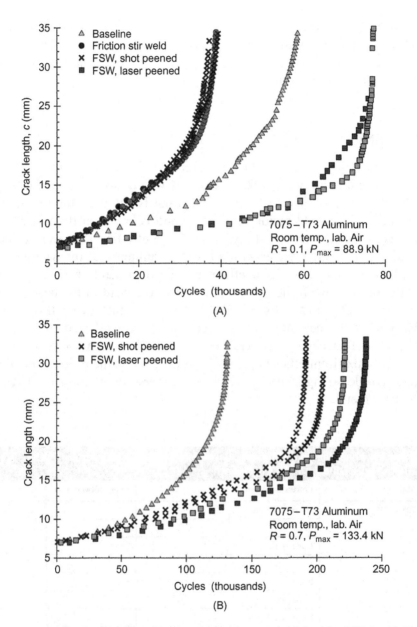

Figure 6.35 Crack length versus number of fatigue cycle in two stress ratios (A) R = 0.1 and (B) R = 0.7 [29].
Source: Reprinted with permission from Springer.

So far, contradicting observations have been reported while comparing the fatigue life of the base metal with the welds. The fatigue behavior was inferior [21] and superior [20] to the base metal, and similar to the base metal. All these observations were made on the 7XXX series Al FSW joints. Furthermore the residual stress state, stress ratio, and the microstructural state also affect the FCG behavior. The origin of these discrepancies needs to be linked to the combination of stress states and microstructure.

6.4 JOINT EFFICIENCY

The joint efficiency is a critical parameter in determining the efficacy of the friction stir welds. There are two widely accepted methods to obtain the joint efficiency of the weld. First is based on the minimum hardness in the weld region and the second is based on the transverse tensile properties. In this study, both the joint efficiencies were calculated to provide the reader with an overall summary of the friction stir welds joint efficiency. Joint efficiency, listed in Table 6.8, is defined as the ratio between the lowest hardness in the weld to the base metal hardness. De and Mishra [30] correlated the heat flux with the joint efficiency of various precipitation-strengthened Al alloys. The joint efficiency of the weld first increased and then decreased with the heat flux. Joint efficiency, listed in Table 6.9, is the ratio of the ultimate tensile strength of the weld in transverse orientation to that of the base metal.

Table 6.8 Minimum HAZ Hardness–Based Joint Efficiency for Various 7XXX Alloys Under Different FSW Conditions			
Material	Process Conditions	Joint Efficiency (%)	References
7050-T7451	150 rpm, 1.7 mm/s		[7]
	In-air	69	
	Under-water	75	
	1000 rpm, 10.2 mm/s		
	In-air	80	
	Under-water	82	
	400 rpm, 5.1 mm/s		
	In-air	78	
	Under-water	81	
7050-T7451	180 rpm, 0.85 mm/s		[17]
	As-weld	55	
	PWHT (121°C for 24 h)	48	

(Continued)

Material	Process Conditions	Joint Efficiency (%)	References
Table 6.8 (Continued)			
7050-T7651	FSW, 700 rpm		[31]
	100 mm/min	67	
	400 mm/min	77	
	SSFSW, 1500 rpm		
	100 mm/min	74	
	400 mm/min	84	
7050-T7451	As-weld		[4]
	90 rpm, 0.85 mm/s	64	
	810 rpm, 3.81 mm/s	77	
	PWHT (120°C, 4 h)		
	90 rpm, 0.85 mm/s	60	
	810 rpm, 3.81 mm/s	70	
7136-T76511	250 rpm, 2.1 mm/s		[15]
	As-weld	56	
	3 years of natural aging	70	
	6 years of natural aging	73	
7050-T7451	540 rpm, 6.8 mm/s		[16]
	FSW pass 1	82	
	FSW pass 2	78	
	FSW pass 3	78	
	FSW pass 4	78	
	FSW pass 5	74	
7010-T7651	140 rpm, 40 mm/min, 40 mm plunge depth	50	[32]
	400 rpm, 150 mm/min, 20 mm plunge depth	60	
7075-T6	1500 rpm		[6]
	40 mm/min	77	
	57 mm/min	80	
	120 mm/min	85	
	170 mm/min	89	
7449 TAF (underaged)	TAF + FSW	46	[18]
	TAF + FSW + PWHT	46	
7010-T7651	450 rpm, 95 mm/min	63	[5]
7075-T7351	Water cooling	61	[13]
	Water cooling + 1000 h natural aging	72	
7050-T7651	200 rpm, 109 mm/min		[14]
	48 h of natural aging	49	
	216 h of natural aging	62	
	36240 h of natural aging	79	

(Continued)

Table 6.8 (Continued)

Material	Process Conditions	Joint Efficiency (%)	References
7075-T651	200 rpm, 109 mm/min		[14]
	48 h of natural aging	48	
	216 h of natural aging	59	
	36,240 h of natural aging	77	
7050	400 rpm, 5.08 mm/s		[9]
	7050-T6 + FSW + T7	69	
	7050-T7 + FSW + T7	70	
	7050-W + FSW + T7	90	
7039-T6	635 rpm		[33]
	As-weld	66	
	Artificial aging (120°C for 18 h)	79	
	1 year of natural aging	83	
	100°C for 8 h +150°C for 24 h	76	

Table 6.9 Transverse Ultimate Tensile Strength–Based Joint Efficiency for Various 7XXX Alloys Under Different FSW Conditions

Material	Process Conditions	Joint Efficiency (%)	References
7075-T651	800 rpm, 100 mm/min	77.8	[34]
	200 mm/min	86.1	
	400 mm/min	91.2	
7136-T76	250 rpm, 2.1 mm/s	71	[35]
7075-T651	12.7 cm/min		[2]
	As-weld	75	
	As-weld + PWHT (121°C for 24 h)	72	
7475 Al	50 mm/min		[36]
	300 rpm	81.6	
	400 rpm	89.5	
	700 rpm	72.8	
	1000 rpm	58	
7039	635 rpm, 75 mm/min		[12]
	FSW joint-T6	85.6	
	FSW joint-W	93.8	
	FSW joint-O	98.2	

(Continued)

Table 6.9 (Continued)

Material	Process Conditions	Joint Efficiency (%)	References
7075-T6	700 rpm, 2.67 mm/s	77	[37]
7050	400 rpm, 5.08 mm/s		[13]
	7050-T6 + FSW	78	
	7050-T7 + FSW	80	
	7050-W + FSW	93	
7075-T6	1250 rpm, 31.5 mm/min		[38]
	As-weld	82	
	Cyclic solution treatment + Aged for 24 h	110	
7039-T6	635 rpm, 75 mm/min		[39]
	As-weld	85.6	
	Air cooling	67.1	
	Liquid nitrogen cooling	69.4	
	Water cooling	73.5	
Al-7B04	800 rpm, 200 mm/min		[10]
	T6-FSW	80	
	T4-FSW	82	
	O-FSW	96	
7075Al-T651	600 rpm, 100 mm/min, butt weld		[40]
	As-weld	75	
	PWHT (T6)	90	
7075Al-T651	250 rpm, 25 mm/min		[41]
	As-weld	70	
	Artificial aging (120°C for 24 h)	56	
	Solution treated + Aging (480°C for 1 h + 120°C for 24 h)	79	
7039-T6	635 rpm		[14]
	As-weld	85.6	
	Artificial aging (120°C for 18 h)	92.2	
	1 year of natural aging	95	
	100°C for 8 h + 150°C for 24 h	80.3	
Al-10Zn-1.9Mg-1.7Cu (in wt.%)	350 rpm, 50 mm/min	70	[42]
	100 mm/min	74	
	150 mm/min	68	
	650 rpm, 100 mm/min	64	
	950 rpm, 100 mm/min	57	

(Continued)

Table 6.9 (Continued)

Material	Process Conditions	Joint Efficiency (%)	References
7039-T6	635 rpm, 75 mm/min	86	[43]
	120 mm/min	85.2	
	190 mm/min	76.6	
	410 rpm, 75 mm/min	66.5	
	540 rpm, 75 mm/min	83.6	
	635 rpm, 75 mm/min	85.6	
7075-O	1000 rpm, 150 mm/min	100.2	[11]
	1500 rpm, 400 mm/min	101.2	
	1000 rpm, 150 mm/min + PWHT (485°C for 4 h + 6 h at 140°C)	93.2	
	1500 rpm, 400 mm/min + PWHT (485°C for 4 h + 6 h at 140°C)	87.5	
7075-T6	1000 rpm, 150 mm/min	79.8	[11]
	1500 rpm, 400 mm/min	67.8	
	1000 rpm, 150 mm/min + PWHT (485°C for 4 h + 6 h at 140°C)	89.1	
	1500 rpm, 400 mm/min + PWHT (485°C for 4 h + 6 h at 140°C)	90.8	
7050-T7451	369 rpm, 1.7 mm/s		[25]
	As-FSW	77	
	As-FSW + T6	75	
	As-FSW + T7	67	
7050-T651	FSW + 5–10 h of natural aging	66	[14]
	FSW + 195 h of natural aging	76	
	FSW + 73,300 h of natural aging	92	

REFERENCES

[1] M.J. Starink, S.C. Wang, A model for the yield strength of overaged Al–Zn–Mg–Cu alloys, Acta Mater. 51 (2003) 5131–5150.

[2] M.W. Mahoney, C.G. Rhodes, J.G. Flintoff, R.A. Spurling, W.H. Bingel, Properties of friction-stir-welded 7075 T651 aluminum, Metall. Mater. Trans. A 29A (1998) 1955–1964.

[3] K.V. Jata, K.K. Sankaran, J.J. Ruschau, Friction-stir welding effects on microstructure and fatigue of aluminum alloy 7050-T7451, Metall. Mater. Trans. A 31A (2000) 2181–2192.

[4] A.P. Reynolds, W. Tang, Z. Khandkar, J.A. Khan, K. Lindner, Relationships between weld parameters, hardness distribution and temperature history in alloy 7050 friction stir welds, Sci. Technol. Weld. Join. 10 (2005) 190–199.

[5] Kh.A.A. Hassan, P.B. Prangnell, A.F. Norman, D.A. Price, S.W. Williams, Effect of welding parameters on nugget zone microstructure and properties in high strength aluminium alloy friction stir welds, Sci. Technol. Weld. Join. 8 (2003) 257–268.

[6] M. Vargas, S. Lathabai, Microstructure and mechanical properties of a friction stir processed Al-Zn-Mg-Cu alloy, Revista Matéria 15 (2012) 270–277.

[7] P. Upadhyay, A.P. Reynolds, Effects of thermal boundary conditions in friction stir welded AA7050-T7 sheets, Mater. Sci. Eng. A 527 (2010) 1537–1543.

[8] R.-D. Fu, Z.-Q. Sun, R.-C. Sun, Y. Li, H.-J. Liu, L. Liu, Improvement of weld temperature distribution and mechanical properties of 7050 aluminum alloy butt joints by submerged friction stir welding, Mater. Des. 32 (2011) 4825–4831.

[9] J. Yan, A.P. Reynolds, Effect of initial base metal temper on mechanical properties in AA7050 friction stir welds, Sci. Technol. Weld. Join. 14 (2009) 582–587.

[10] Y. Chen, H. Ding, Z. Cai, J. Zhao, J. Li, Effect of initial base metal temper on microstructure and mechanical properties of friction stir processed A1-7B04 alloy, Mater. Sci. Eng. A 650 (2016) 396–403.

[11] G. İpekoğlu, S. Erim, G. Cam, Effects of temper condition and post weld heat treatment on the microstructure and mechanical properties of friction stir butt-welded AA7075 Al alloy plates, Int. J. Adv. Manuf. Technol. 70 (2014) 201–213.

[12] C. Sharma, D.K. Dwivedi, P. Kumar, Influence of pre-weld temper conditions of base metal on microstructure and mechanical properties of friction stir weld joints of Al–Zn–Mg alloy AA7039, Mater. Sci. Eng. A 620 (2015) 107–119.

[13] T.W. Nelson, R.J. Steel, W.J. Arbegast, In situ thermal studies and post-weld mechanical properties of friction stir welds in age hardenable aluminium alloys, Sci. Technol. Weld. Join. 8 (2003) 283–288.

[14] C.B. Fuller, M.W. Mahoney, M. Calabrese, L. Micona, Evolution of microstructure and mechanical properties in naturally aged 7050 and 7075 Al friction stir welds, Mater. Sci. Eng. A 527 (2010) 2233–2240.

[15] I. Kalemba, C. Hamilton, S. Dymek, Natural aging in friction stir welded 7136-T76 aluminum alloy, Mater. Des. 60 (2014) 295–301.

[16] R. Brown, W. Tang, A.P. Reynolds, Multi-pass friction stir welding in alloy 7050-T7451—effects on weld response variables and on weld properties, Mater. Sci. Eng. A 513–514 (2009) 115–121.

[17] C.T. Canaday, M.A. Moore, W. Tang, A.P. Reynolds, Through thickness property variations in a thick plate AA7050 friction stir welded joint, Mater. Sci. Eng. A 559 (2013) 678–682.

[18] A. Sullivan, J.D. Robson, Microstructural properties of friction stir welded and post-weld heat-treated 7449 aluminium alloy thick plate, Mater. Sci. Eng. A 478 (2008) 351–360.

[19] Fatigue book reference.

[20] C. Sharma, D.K. Dwivedi, P. Kumar, Fatigue behavior of friction stir weld joints of Al–Zn–Mg alloy AA7039 developed using base metal in different temper condition, Mater. Des. 64 (2014) 334–344.

[21] M. Kadlec, R. Růžek, L. Nováková, Mechanical behaviour of AA 7475 friction stir welds with the kissing bond defect, Int. J. Fat. 74 (2015) 7–19.

[22] S. Di, X. Yang, D. Fang, G. Luan, The influence of zigzag-curve defect on the fatigue properties of friction stir welds in 7075-T6 Al alloy, Mater. Chem. Phy. 104 (2007) 244–248.

[23] A.H. Feng, D.L. Chen, Z.Y. Ma, Microstructure and cyclic deformation behavior of a friction-stir-welded 7075 Al alloy, Metall. Mater. Trans. 41A (2010) 957–971.

[24] C. Deng, H. Wang, B. Gong, X. Li, Z. Lei, Effects of microstructural heterogeneity on very high cycle fatigue properties of 7050-T7451 aluminum alloy friction stir butt welds, Int. J. Fat. 83 (2016) 100–108.

[25] R. John, K.V. Jata, K. Sadananda, Residual stress effects on near-threshold fatigue crack growth in friction stir welds in aerospace alloys, Int. J. Fat. 25 (2003) 939–948.

[26] Y. Uematsu, K. Tokaji, Y. Tozaki, H. Shibata, Fatigue behavior of friction stir welded A7075-T6 aluminium alloy in air and 3% NaCl solution, Weld. J. 27 (2013) 441–449.

[27] Nilesh's residual stress book.

[28] O. Hatamleh, J. Lyons, R. Forman, Laser and shot peening effects on fatigue crack growth in friction stir welded 7075-T7351 aluminum alloy joints, Int. J. Fat. 29 (2007) 421–434.

[29] O. Hatamleh, S. Forth, A.P. Reynolds, Fatigue crack growth of peened friction stir-welded 7075 aluminum alloy under different load ratios, J. Mater. Eng. Perform. 19 (2010) 99–106.

[30] P.S. De, R.S. Mishra, Friction stir welding of precipitation strengthened aluminium alloys: scope and challenges, Sci. Technol. Weld. Join. 16 (2011) 343–347.

[31] H. Wu, Y.-C. Chen, D. Strong, P. Prangnell, Stationary shoulder FSW for joining high strength aluminum alloys, J. Mater. Process. Technol. 221 (2015) 187–196.

[32] A. Sullivan, C. Derry, J.D. Robson, I. Horsfall, P.B. Prangnell, Microstructure simulation and ballistic behavior of weld zones in friction stir welds in high strength aluminium 7xxx plate, Mater. Sci. Eng. A 528 (2011) 3409–3422.

[33] C. Sharma, D.K. Dwivedi, P. Kumar, Effect of post weld heat treatments on microstructure and mechanical properties of friction stir welded joints of Al–Zn–Mg alloy AA7039, Mater. Des. 43 (2013) 134–143.

[34] A.H. Feng, D.L. Chen, Z.Y. Ma, W.Y. Ma, R.J. Song, Microstructure and strain hardening of a friction stir welded high-strength Al-Zn-Mg alloy, Acta Metall. Sin. 27 (2014) 723–729.

[35] I. Kalemba, S. Dymek, C. Hamilton, M. Blicharski, Microstructure and mechanical properties of friction stir welded 7136–T76 aluminium alloy, Mater. Sci. Technol. 27 (2011) 903–908.

[36] R.K. Gupta, H. Das, T.K. Pal, Influence of processing parameters on induced energy, mechanical and corrosion properties of FSW butt joint of 7475 AA, J. Mater. Eng. Perform. 21 (2012) 1645–1654.

[37] P. Cavaliere, A. Squillace, High temperature deformation of friction stir processed 7075 aluminium alloy, Mater. Charact. 55 (2005) 136–142.

[38] S.M. Bayazid, H. Farhangi, H. Asgharzadeh, L. Radan, A. Ghahramani, A. Mirhaji, Effect of cyclic solution treatment on microstructure and mechanical properties of friction stir welded 7075 Al alloy, Mater. Sci. Eng. A 649 (2016) 293–300.

[39] C. Sharma, D.K. Dwivedi, P. Kumar, Influence of in-process cooling on tensile behavior of friction stir welded joints of AA7039, Mater. Sci. Eng. A 556 (2012) 479–487.

[40] S.R. Ren, Z.Y. Ma, L.Q. Chen, Effect of initial butt surface on tensile properties and fracture behavior of friction stir welded Al–Zn–Mg–Cu alloy, Mater. Sci. Eng. A 479 (2008) 293–299.

[41] P. Sivaraj, D. Kanagarajan, V. Balasubramanian, Effect of post weld heat treatment on tensile properties and microstructure characteristics of friction stir welded armor grade AA7075-T651 aluminium alloy, Def. Technol. 10 (2014) 1–8.

[42] F. Zhang, X. Su, Z. Chen, Z. Nie, Effect of welding parameters on microstructure and mechanical properties of friction stir welded joints of a super high strength Al–Zn–Mg–Cu aluminum alloy, Mater. Des. 67 (2015) 483–491.

[43] C. Sharma, D.K. Dwivedi, P. Kumar, Effect of welding parameters on microstructure and mechanical properties of friction stir welded joints of AA7039 aluminum alloy, Mater. Des. 36 (2012) 379–390.

CHAPTER 7

Corrosion

7.1 INTRODUCTION

Apart from the mechanical properties of the friction stir welding (FSW) joints, corrosion is a critical property as the structural integrity of the components deteriorates with time as the environmental damage initiates and propagates. The following are the major forms of corrosion that attacks various zones of the FSW joints to varying extent: exfoliation corrosion, galvanic corrosion, intergranular corrosion, and stress corrosion cracking. The localized corrosion susceptibility in 7XXX Al alloys is related to the presence of coarse grain boundary precipitates and wide precipitate-free zone. Localized attacks in the form of pits first occur at or around the grain boundary precipitates or the precipitate-free zones and the damage grows as intergranular corrosion. And in the case of structural components, when a material that is susceptible to intergranular corrosion is used under load, then the problem of the stress corrosion cracking emerges.

The corrosion resistance of the weld zones in as-welded condition is inferior to the base metal [1]. Moreover, 7XXX alloys continue to naturally age over a very long time and as was discussed in Chapter 6, Mechanical Properties, it could be longer than 7 years. Therefore the corrosion susceptibility, similar to the strength, fatigue, and toughness evolution, also changes over the years, more specifically, it deteriorates. The most feasible solution to stabilize the response is to postweld heat treat the weld to restore the corrosion resistance. Most of the heat treatment conditions, including 121°C for 24 h and 100°C for 100 h, were proven to be effective in improving the corrosion resistance of the FSW joints [1]. In addition the base metal temper plays a critical role in controlling the corrosion response of the weld joints. Widener [1] conducted a detailed investigation on the role of initial temper (T6 and T7) and also on the effect of various postweld heat treatment procedure on the corrosion response of the 7075 Al joints [1]. The following are the usual

Friction Stir Welding of High-Strength 7XXX Aluminum Alloys. DOI: http://dx.doi.org/10.1016/B978-0-12-809465-5.00007-6

exfoliation ratings description: N, no apparent attack (0 points); EA, superficial attack (1 points); EB, moderate attack (2 points); EC, severe attack (3 points); and ED, very severe attack (4 points). The exfoliation corrosion test results of the 7075-T73 weld zones after various postweld heat treatments are shown in Fig. 7.1. The results of the exfoliation corrosion ratings for 7075-T73 weld are summarized in Table 7.1. Best exfoliation corrosion resistance was observed for the following heat treatment conditions: 163°C for 2, 4, and 8 h, and 100°C for 100 h. Note that the corrosion in the naturally aged weld was severe as compared to any other heat treatment conditions. The exfoliation corrosion test results under various postweld heat treatment conditions on the 7075-T6 joints are given in Table 7.2. The treatments that resulted in superior corrosion resistance in the case of the 7075-T73 joints did actually deteriorate the exfoliation corrosion resistance in the case of the 7075-T6 weld. The retrogression re-aging treatment resulted in better corrosion properties as compared to other heat treatment conditions. By comparing the corrosion response the 7075-T73 condition exhibited superior response than the 7075-T6 even after various heat treatment conditions optimization. Hence the role of not only the initial base temper but also the subsequent postweld heat treatment is important for corrosion. There was no correlation between the corrosion responses of various zones after various postweld heat treatments with the

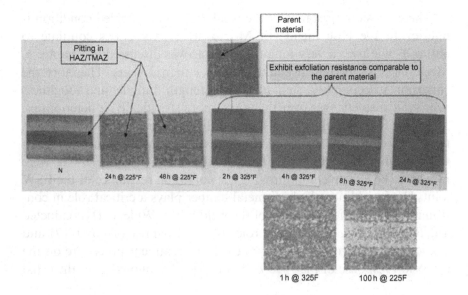

Figure 7.1 Exfoliation test results after various postweld heat treatments [1].

observed microstructural evolution [1]. However, based on the knowledge and information presented in Chapters 5 and 6, link between the initial temper, welding parameters, heat treatments, microstructural evolution, and mechanical properties is existed. Therefore, to have a complete understanding of the corrosion behavior, a combination of characterization techniques have to be employed which is discussed at the end of this chapter.

The example of the initial stages of the intergranular corrosion and the preferential attack of the grain boundary precipitates and

Table 7.1 Summary of the Exfoliation Test Ratings After Various Postweld Heat Treatments in 7075-T73 Weld [1]

Alloy	PWAA	Weld Zone	HAZ	Base Material
7075-T73	Parent			EA/EB
	N	EC	ED	EA/EB
	24 h @ 225F	EB	EC	EA/EB
	48 h @ 225F	EA	EB	EA/EB
	100 h @ 225F	EA	EA	EA/EB
	1 h @ 325F	EA	EA/EB	EA/EB
	2 h @ 325F	N	EA	EB
	4 h @ 325F	N	EA	EB
	8 h @ 325F	N	EA	EB
	24 h @ 325F	EB	EA	EB

Table 7.2 Summary of the Exfoliation Test Ratings After Various Postweld Heat Treatments in 7075-T6 Weld [1]

Alloy	PWAA	Weld Zone	HAZ	Base Material
7075-T6	Parent			EA/EB
	N	EC	ED	EA
	24 h @ 225F	EC	ED	EA
	100 h @ 225F	ED	EC	EA
	8 h @ 320F + 24 h @ 250F	P/EA	ED	P/ED
	11 h @ 320F + 24 h @ 250F	EA	ED	P/EC
	2 h @ 355F + 24 h @ 250F	N/EA	ED	EA/EB
	3 h @ 355F + 24 h @ 250F	N/EA	EB/EC	EA/EB
	26 h @ 325F	EA/EB	EC	EA/EB
	9 h @ 355F	EA/EB	EC	EA/EB
	12 h @ 375F	EB	ED	EB

precipitate-free zones are shown in Fig. 7.2 [2]. Note that the grain interior shows no impact due to the presence of the corrosive environment. The heat-affected zone (HAZ) and the nugget response to the immersion tests (ASTM G110-92) can be observed in Fig. 7.2A and B, respectively. The black region is the corroded precipitate-free zone and the white region are the remaining grain boundary precipitates.

The response of the thermomechanically affected zone (TMAZ) and HAZ after the immersion test in a 3.5 wt.% NaCl solution (ASTM G129) is shown in Fig. 7.3A and B, respectively. In the TMAZ the corrosion was more pit-like, and in the heat-affected zone a more severe intergranular type corrosion was observed [2].

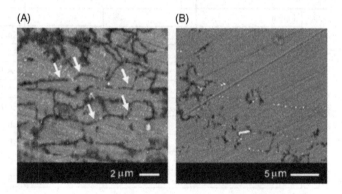

Figure 7.2 Immersion study to observe the distinct response of the (A) heat-affected zone and (B) nugget regions to the corrosive environment (Images acquired in BSE mode) [2]. Source: Reprinted with permission from Elsevier.

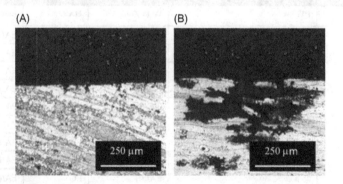

Figure 7.3 Corrosion response to 3 wt.% NaCl solution in (A) thermomechanically affected zone and (B) heat-affected zone [2]. Source: Reprinted with permission from Elsevier.

Furthermore, sensitization of 7050 FSW joint takes place in the weld zones in the form of Cu-Zn-rich precipitates. The sensitization in TMAZ in the case of 7050-T7451 weld is shown in Fig. 7.4. Grain boundary sensitization also promotes intergranular corrosion.

Paglia et al. [3] investigated the localized corrosion response and stress corrosion cracking (SCC) susceptibility in 7075-T651 and 7050-T7451 FSW joints made under the same welding parameters. Both the welds exhibited the typical W-shaped profile, and the nugget strength of the 7050-T7451 was higher than the 7075-T651 weld. Potentiodynamic scans, immersion testing, and SCC tests were carried out for both the welds and the following conclusions were made. The results are shown in Figs. 7.5 and 7.6. In 7075-T651 FSW the pitting potential was lower in HAZ, large corrosion products were observed in HAZ, and the SCC failure occurred along HAZ. Therefore the corrosion susceptibility of the HAZ region

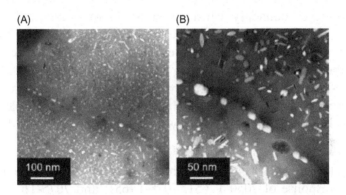

Figure 7.4 Grain boundary sensitization in the thermomechanically affected region of a 7050-T7451 weld [2].
Source: Reprinted with permission from Elsevier.

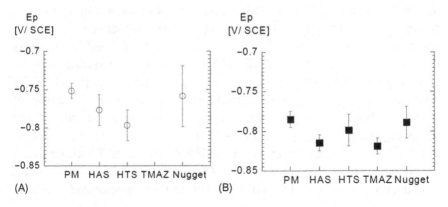

Figure 7.5 Pitting potentials in various zones of (A) 7075-T651 and (B) 7050-T7451 FSW [3].
Source: Copyright 2003 by The Minerals, Metals & Materials Society. Reprinted with permission.

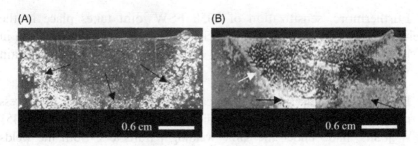

Figure 7.6 Observation of the corrosion products after the immersion testing in (A) 7075-T651 and (B) 7050-T7451 FSW [3]. Source: Copyright 2003 by The Minerals, Metals & Materials Society. Reprinted with permission.

was highest as compared to the other regions. This was attributed to the Cu depletion from the HAZ grain boundaries. In the case of 7050-T7451 FSW pitting potential was lower in TMAZ and corrosion products in large quantity was observed in TMAZ and also at the bottom of the nugget. The SCC failure occurred along HAZ when tested in air and along TMAZ−nugget boundary when tested in 3.5 wt.% NaCl. The grain boundary chemical mapping of the TMAZ boundaries revealed the presence of Cu-enriched $Mg(Zn,Cu)_2$ precipitates and this could be the reason for the observed localized attack along the TMAZ grain boundaries. The 7075-T651 TMAZ was relatively free of grain boundary precipitates as opposed to the 7050-T7541 TMAZ boundaries. A possible reason that the TMAZ grain boundaries were relatively free of the localized intergranular attack.

Lumsden et al. [4] compared the immersion and stress corrosion cracking response of 7050-T7651, 7075-T7651, and 7075-T651. In the case of 7050-T7651 the intergranular attack during immersion testing started at the nugget−TMAZ boundary and spread into the nugget region. In addition, lowest pitting potential, stress corrosion cracking susceptibility, and failure were observed in the nugget region. On the other hand, in the case of 7075 Al alloys in both T6 and T7 tempers the attack started at the TMAZ−HAZ boundary and spread into the HAZ region. The pitting potential was also the lowest in the HAZ for both the tempers. In another study on the corrosion behavior of 7075-T6, Paglia et al. [5] observed the corrosion resistance of the retreating side HAZ to be inferior to the remaining weld zones. Particularly the lowest potential at the HAZ and the SCC failure along the HAZ, which was due to the Cu depletion along these grain boundaries. Both 7075-T6 and 7075-T76 tempers responded similarly to the corrosion susceptibility. In the case of 7108-T79 FSW joints severe corrosion attack during exfoliation immersion corrosion occurred at the edges of the TMAZ [6]. The localized attack was intergranular within the TMAZ and extended to the HAZ region. Furthermore, open circuit potential measurements resulted in the lowest potential along the TMAZ region which signifies the corrosion susceptibility. TEM characterization revealed the presence of intergranular corrosion. The behavior was attributed to the presence and absence of $MgZn_2$ precipitates along the grain boundaries and in the grain interior, respectively, and a resultant potential difference that favored the anodic corrosion activity along the grain boundaries. Based on these studies, corrosion behavior of different FS weld zones are highly sensitive to the alloy composition rather to the initial base temper of the alloy. Note that both Cu and Zn contents are higher in 7050 Al as compared to the 7075 Al alloy.

The effect of temper and artificial aging on the corrosion susceptibility of 7075 Al alloys was investigated by Paglia et al. [7]. The 7075-O and 7075-T7451 Al alloys were friction stir welded, and the 7075-O FSW condition was subjected to subsequent T7451 over-aging heat treatment. For all the three tempers lowest hardness was observed at the HAZ. As expected the transverse tensile samples failed within the HAZ. The following microstructural features were observed. In 7075-O temper, less intragranular precipitates, grain boundary phases, and a small precipitate-free zone were observed in the HAZ as compared to the T7451 overaged condition. 7075-O + FSW + T7451 condition also exhibited a narrow precipitate-free zone with intra-granular precipitates. The 7075-T7451 material exhibited coarse precipitates along with the precipitate-free zone as compared to the other two conditions. The effect of these different precipitate sizes can be distinctly observed on the corrosion behavior. The ASTM G 110-92 immersion corrosion test results are shown in Fig. 7.7. As can be clearly seen, severe corrosion attack along the thermomechanically affected and HAZs were observed in 7075-T7451 FSW followed by the 7075-O FSW and a reduced attack in 70754-O FSW + T7451 treatment. A severe localization in the case of in 7075-T7451 FSW was due to the severe precipitate coarsening and more pronounced precipitate-free zone.

In 7050-T7451 FSW a higher ΔK_{th} and lower crack growth rates were found in the HAZ as compared to the base metal and nugget

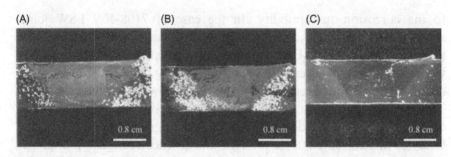

Figure 7.7 Transverse cross section observation after the immersion testing in (A) 7075-O FSW, (B) 7075-T7 FSW, and (C) T7 postweld heat-treated 7075-O FSW [7]. Source: Reprinted with permission from Wiley and Sons.

Table 7.3 A Detailed Summary of the Microstructural Features at Various Zones, the Expected Corrosion Response, and the Combination of the Appropriate to Technique to Characterize the Localized Corrosion Response

Zone	Microstructure	Localized Corrosion	Localized Corrosion Investigation Technique
Parent metal	Strengthening precipitates	Pitting	Polarization techniques, spray tests droplet cell methods, and other methods
	Small grain boundary phases	Intergranular corrosion	
	Narrow precipitate-free zones	Other corrosion phenomena	
Heat-affected zone (HAZ)	Coarse intragranular precipitates	Intergranular corrosion	Corrosion immersion tests: appropriate for the investigation of intergranular corrosion if combined with microstructure investigations (optical or SEM in backscattered mode)
	Coarse grain boundary phases	Intersubgranular corrosion	Polarization techniques: not appropriate for a clear discrimination of the extent of the intergranular corrosion
	Wide precipitate-free zones	Pitting	
Thermomechanically affected zone	Variable dimensions of the precipitates and grain boundary phases	Intergranular corrosion	Similar as for the HAZs
Nugget	General absence of intragranular precipitates and precipitate-free zones	Pitting	Immersion, polarization tests with microstructural investigations (optical or SEM in backscattered mode)
	For some alloys, the presence of intragranular precipitates and precipitate-free zones	Intergranular corrosion	Conventional methods appropriate for pitting corrosion

Source: *Directly adopted from C.S. Paglia, R.G. Buchheit, A look in the corrosion of aluminum alloy friction stir welds, Scr. Mater. 58 (2008) 383–387.*

when tested in air and in 3.5 wt.% NaCl solution [8]. At intermediate and higher ΔK values, for all three tested zones, the crack growth rates under NaCl solution were higher than in air. A possible explanation for the lower crack growth rates in the HAZ is be due to the compressive residual stress that promotes the crack closure under both the tested environments. Note that the residual stress state dominated the corrosion–fatigue crack growth behavior but not the severely coarsened precipitates, which otherwise would have led to a faster damage accumulation and failure (Table 7.3).

Overall, both the alloy chemistry and the subsequent postweld heat treatment dominated the corrosion susceptibility of various zones. There is no variation between T6 and T7 temper corrosion response. Furthermore, both the intergranular and intragranular precipitate characteristics predominantly determined the corrosion response of the different weld zones.

REFERENCES

[1] C. Widener, Evaluation of post-weld heat treatments for corrosion protection in friction stir welded 2024 and 7075 aluminum alloys (Ph.D. dissertation), Wichita State University, 2005.

[2] C.S. Paglia, R.G. Buchheit, A look in the corrosion of aluminum alloy friction stir welds, Scr. Mater. 58 (2008) 383–387.

[3] C.S. Paglia, L.M. Ungaro, B.C. Pitts, M.C. Carroll, A.P. Reynolds, R.G. Buchheit, The corrosion and environmentally assisted cracking behavior of high strength aluminum alloys friction stir welds: 7075-T651 vs. 7050-T7451, in: K.V. Jata, M.W. Mahoney, R.S. Mishra, S.L. Semiatin, T. Lienert (Eds.), Friction Stir Welding and Processing II, TMS, Warrendale, PA, 2003, p. 65.

[4] J.B. Lumsden, M.W. Mahoney, G.A. Pollock, Corrosion behavior of friction stir welded high strength aluminum alloys, This paper is part of the following report: Tri-Service Corrosion Conference, ADA426685.

[5] C.S. Paglia, M.C. Carroll, B.C. Pitts, A.P. Reynolds, R.G. Buchheit, Strength, corrosion and environmentally assisted cracking of a 7075- T6 friction stir weld, Mater. Sci. Forum 396–402 (2002) 1677–1684.

[6] D.A. Wadeson, X. Zhou, G.E. Thompson, P. Skeldon, L. Djapic Oosterkamp, G. Scamans, Corrosion behaviour of friction stir welded AA7108 T79 aluminium alloy, Corr. Sci. 48 (2006) 887–897.

[7] C.S. Paglia, K.V. Jata, R.G. Buchheit, The influence of artificial aging on the microstructure, mechanical properties, corrosion, and environmental cracking susceptibility of a 7075 friction-stir-weld, Mater. Corr. 58 (2010) 737–750.

[8] P.S. Pao, S.J. Gill, C.R. Feng, K.K. Sankaran, Corrosion–fatigue crack growth in friction stir welded Al 7050, Scr. Mater. 45 (2001) 605–612.

CHAPTER 8

Physical Metallurgy-Based Guidelines for Obtaining High Joint Efficiency

In Chapter 6, Mechanical Properties, we presented a compilation of joint efficiency values for various alloys in different tempers and postweld heat treatment (PWHT) conditions. It can be noted that the broader approach is to select PWHT approach that replicates the conventional step. For example, it is common to use the T6 aging step. But if we consider the fundamental behind the T6 aging temperature and time, it is based on a specific alloy and the efficacy depends on supersaturation of solutes as well as vacancy concentration from the preceding solid solution step. The nugget, thermomechanically affected zone, and heat-affected zone (HAZ) are in different microstructural states based on the process parameters and thermal boundary conditions. Based on the information in Chapter 4, Temperature Distribution; Chapter 5, Microstructural Evolution; and Chapter 6, Mechanical Properties, this is a short discussion to outline some guidelines and as well as highlight some potential future approaches for high joint efficiency.

There are three major design-related variations in terms of alloy selection, starting temper, and thickness of material. Microstructurally the most critical aspect is the types of final precipitates. Artificial aging for T6 and T7 tempers leads to η' and η precipitates. For conventional applications, these are the final precipitates. Fig. 8.1 shows a schematic to capture the scaling of tool size with the thickness of material. As mentioned in previous chapters, multiple things happen as the combination of material thickness and tool dimension changes. First, the heat input from tool increases with the thickness of material as the tool traverse rate becomes lower. Second, the heat extraction becomes lower with increase in the thickness of workpiece. The impact is on the width and angle of the HAZ plane. As the fracture path during transverse loading matches the angle of HAZ plane, this impacts the final performance. A simple estimate of the angle of fracture path can be calculated as $\tan \frac{t}{W_{\text{HAZ}}}$. Note that this brings in the intersection of mechanics and

Friction Stir Welding of High-Strength 7XXX Aluminum Alloys. DOI: http://dx.doi.org/10.1016/B978-0-12-809465-5.00008-8

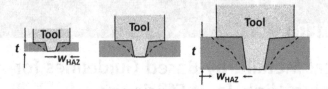

Figure 8.1 An illustration of scaling of tool with the thickness of material. It changes the angle of HAZ and the associated fracture path during transverse loading.

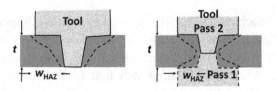

Figure 8.2 An illustration of single pass versus double pass approach for thick plates. Note the change in the overall HAZ path in the double pass weld. The associated fracture path during transverse loading will change as the crack path becomes more complicated for the double pass weld.

metallurgical factors leading to overall performance of joint. This aspect can be appreciated further by consideration of friction stir welding of thick plates by single pass or double pass approach. Fig. 8.2 shows the change in the overall HAZ pattern after the double pass weld. Note that the traverse rate can be higher for double pass weld because of the smaller tool. That again lowers the width of the HAZ region. The combined effect would be higher performance in double pass weld. Of course, from the design perspective the accessibility of the plate for double side weld is required for such approach. It is important to note that this is not similar to the double side bobbin tool approach. The use of bobbin tool eliminates the backing plate that extracts heat more efficiently. The impact of these approaches on the distribution of precipitates and PWHT response is quite significant.

For 7XXX alloys, the use of differential scanning calorimetry (DSC) in the optimization of tool rotation rate and traverse rate can be very effective. If PWHT is an option for the structure being welded, then it is important to have enough solute in solid solution in the critical region for strength recovery. The DSC signature can guide the PWHT approach. So far, the literature is available for very long-term natural aging and conventional artificial aging. A key gap is the low-temperature artificial aging response. Additionally, scientific understanding of the alloy chemistry on PWHT is needed. Conceptually, two separate aspects are important: enhanced solute retention after the process and high nucleation density during the PWHT.

CHAPTER 9

Summary and Future Outlook

Joining of the high-strength aluminum alloys via solid-state friction stir welding (FSW) has been exceedingly successful. A great extent of knowledge exists on the role of various FSW process parameters in obtaining defect-free, high-quality welds, on the formation of defects and banded heterogeneous microstructure, and finally in the creation of the process map to guide the future endeavors.

Temperature variation across different weld zones is inherent to the welding techniques, including the relatively low peak temperature in solid-state friction stir welds. As a result, there is a variation in both the microstructural features and the corresponding mechanical properties. Therefore an appropriate control of the thermal fields during and/or after the welding with the help of intrinsic process parameters and/or external heating or the cooling media greatly modifies the results. The thermal fields have to be applied based on good understanding of microstructural evolution in various zones and also on the expected nature of evolution. For example, different approach is needed to avoid precipitate coarsening during the thermal cycle or reprecipitation during the cooling cycle. As compared to the base metal, each zone exhibits an extremely heterogeneous microstructure and the level of solute supersaturation. Therefore care should be taken in selection of the postweld heat treatment procedures. It should be based on the physical metallurgy applicable to the particular zone of interest. Most critical aspect is the development of new heat treatment procedures by considering the microstructural heterogeneity and variation. Particularly in order to achieve a high joint efficiency structure.

The transverse mechanical properties completely reflect the variation in the microstructure across the weld nugget. In most cases lowest hardness points are observed at the heat-affected zone (HAZ) or at the HAZ—thermomechanically affected zone (TMAZ) boundary, the location that corresponds to severe precipitate coarsening. As expected, the plastic strain localization and eventual sample failure occur along

Friction Stir Welding of High-Strength 7XXX Aluminum Alloys. DOI: http://dx.doi.org/10.1016/B978-0-12-809465-5.00009-X

the lowest hardness region. Transverse-weld specimen elongation deteriorates due to the localization which can be avoided by having smaller difference between the minima and maxima, and also by manipulating the width of the lowest hardness region. Research is needed on the mechanics of flow localization and its dependence on the geometrical nature of strength gradient. Some key results have shown that long-term natural aging of the weld leads to almost full recovery of the strength across the weld regions. This opens up possibilities of low-temperature artificial aging. While strength recovery is attained, any negative impact on the corrosion properties needs to be avoided.

The maximum joint efficiency in T6 or T7 base tempers has reached around 90%, which is by far better than any other welding techniques. It also depends on the thickness of material. The values are quite low for thicker material. Obviously the improvement in the overall structural efficiency toward 100% joint efficiency will be enormous. Based on the current understanding of the FSW process, both the processing parameters and the postwelding treatments conditions have to be systematically investigated to get $\sim$ 100% joint efficiency. Results on alloys with O temper have shown close to 100% joint efficiency. This clearly suggests a direction where the FSW needs to be performed in O or W temper. It is important to note that these tempers are very unstable and age rapidly at ambient temperature. So, to effectively use this approach, the FSW step will need to be integrated at primary metal producers and the welding has to be done as soon as the material in these tempers are produced. The current manufacturing paradigm involves alloy production by primary metal producer, design by original equipment manufacturers (OEMs) and fabrication by the companies in OEM's supplier chain. To take full advantage of using O or W temper material, this manufacturing paradigm will need to change.

Furthermore the composition of the alloys can be adjusted to introduce microstructural features that are more stable compared to the conventional 7XXX alloys. This can be a long-term approach to fully exploit the attributes of FSW. The current 7XXX alloys include many alloys which are specifically developed for specific applications. For example, the alloy chemistry is optimized for thick plate applications where the thermomechanical processing strain is low. Or the alloy composition for extruded panels is different from the rolled products. Similar effort toward alloys designed for FSW will be groundbreaking and provide significant flexibility to designers.